安住

杨梅竹斜街改造纪实与背后的思考

HOME

中国公共艺术丛书 1
PUBLIC ART IN CHINA SERIES

HOME – Documentaries and Thoughts Behind the Reconstruction Project of Yangmeizhu Xiejie

文化艺术出版社
Culture and Art Publishing House

童岩 黄海涛 谢晓英 / 著
Tongyan Huanghaitao Xiexiaoying

图书在版编目（CIP）数据

安住·杨梅竹斜街改造纪实与背后的思考 / 童岩，黄海涛，谢晓英著. — 北京：文化艺术出版社，2018.3（2024.8重印）
ISBN 978-7-5039-6468-8

Ⅰ.①安… Ⅱ.①童… ②黄… ③谢… Ⅲ.①城市道路-城市规划-研究 Ⅳ.① TU984.191

中国版本图书馆CIP数据核字（2018）第043030号

安住·杨梅竹斜街改造纪实与背后的思考

著　　者	童　岩　黄海涛　谢晓英
责任编辑	吴士新　李　特
书籍设计	黄海涛　楚燕平　高博翰
翻　　译	Wang Yile（澳大利亚）　Daniel Lenk（英国）　杨　帆
出版发行	文化藝術出版社
地　　址	北京市东城区东四八条52号　（100700）
网　　址	www.caaph.com
电子邮箱	s@caaph.com
电　　话	（010）84057666（总编室）　84057667（办公室）
	84057696—84057699（发行部）
传　　真	（010）84057660（总编室）　84057670（办公室）
	84057690（发行部）
经　　销	新华书店
印　　刷	北京金彩印刷有限公司
版　　次	2019年2月第1版
印　　次	2024年8月第2次印刷
开　　本	700毫米×1000毫米　1/16
印　　张	21.75
字　　数	182千字
书　　号	ISBN 978-7-5039-6468-8
定　　价	100.00 元

版权所有，侵权必究。如有印装错误，随时调换。

中国公共艺术丛书编辑委员（按姓氏笔画为序）

王　中　王永刚　王明贤　马钦忠
孙振华　汪大伟　吴士新　杨奇瑞
翁剑青　焦兴涛　景育民

序一　前线的"胡同花草堂"

王明贤

建筑与艺术史学者
第十届威尼斯建筑双年展中国国家馆策展人

杨梅竹斜街环境改造和持续的社区营造，是北京老城有机更新一个重要的实验性项目。

在第15届威尼斯国际建筑双年展上，由中国城市建设研究院无界景观工作室关于杨梅竹斜街66—76号院夹道景观改造而延伸出的综合装置作品"安住·平民花园"在威尼斯国际建筑双年展中国馆展出。作品从设计师和普通百姓之间的对话这个特殊的角度去反映中国的城市问题，在威尼斯建筑双年展这个国际建筑最重要的展示空间上和国际建筑界进行直接交流。它出类拔萃、与众不同，引发人了们的各种思考。

本届威尼斯建筑双年展以"前线报告"(Reporting from

the Front)为总主题,从"仿生学和未来""难民和战争""非正式结构""自然和地缘政治""手工艺和传统""家和归属感"六大主题阐释怎样以当代建筑来处理经济不平等、资源分配不均、能源消耗、自然灾害和公共空间等问题的方法论。而本届威尼斯国际建筑双年展中国馆策展的理念是"平民设计,日用即道",这是对双年展总主题做出的思考和回应,更是当下中国社会现象和文化问题在建筑方面的真实反映。"安住·平民花园"深层的社会关怀理念和中国馆的策展理念不谋而合,在学术上呼应了这一主题。这次在威尼斯城处女花园的现场搭建,体现了中国设计师和百姓面对中国城市现状的另一种思想态度和工作方式。

威尼斯双年展是国际最著名的当代艺术大展,堪称艺术界的奥林匹克。作为双年展这一特殊展览形式,它的每一次展览都反映了当代艺术最前沿的状态,预示着当代艺术发展的潮流方向。它所包含的国际视觉艺术双年展和建筑双年展分单双年轮流举行,威尼斯建筑双年展1980年首次举行,在30多年的历程中成为世界建筑艺术和学术界最具影响力的盛事,是目前世界顶级的建筑艺术双年展,对世界建筑发展的方向有举足轻重的影响。

在中国,到处都在进行大规模的建设、拆迁、改造,一片混乱。"安住·平民花园"考虑到现在的中国城市发展以及如何让普通百姓参与到当代建筑中来，通过展示在6个月展期中不断生长变化的装置，参观者可参与植物的播种与采摘，来体验人与设计、人与人的互动。装置内部还设有直播设备，让威尼斯的参观者可看到北京杨梅竹斜街的实况录像，很多细节都会触动心房。它的成功展出引起了国际建筑界艺术界的广泛关注。

在高速城市化的进程中，北京的老胡同不断消失，北京大栅栏片区的杨梅竹斜街这少有的历史碎片能存留下来实属不易。2012年开始的杨梅竹斜街环境改造工程是一个多层次的协作项目，以不改变胡同肌理为前提，强调新生活元素、新商业业态和居民原有的生活和谐共生。无界景观工作室在项目实施后进行调研，发现原住民对于胡同改造的态度以及居民自发种植的行为，因此在后期工作中的角色发生了转变，从为居民设计到引导居民自发营造。

2015年，无界景观工作室开始对杨梅竹斜街66—76号院夹道进行社区改造，以建立"胡同花草堂"为中介，为常住或暂居于此的五户居民建立有效的邻里交往方式，通过养花、种菜等形式相互交流，创造社区共享价值，甚至让暂住人口也能通过花草堂的种植找到归属感，提高了胡同居住者的生活质量。无界景观工作室主持人谢晓英认为，"我们期待的目标是，在保存该城区原有肌理的条件下升级及改造，使当地居民从'不安'到'安住'，由此建立衰败街区与当代城市生活的接口"。

无界景观工作室先是修整铺装、增建无障碍设施、拓宽夹道，使胡同居民的公共生活环境得到改善。接着，以建立共享花草堂的方式介入社区营造，让居住者通过养花、种菜等自然中介的形式互相交流，促进邻里关系的良性发展。专业设计师苦苦探索的景观美、园艺美，对于生活在底层的百姓来说并没有那么重要。老百姓更愿意在家门口的种植池中种一些小葱、丝瓜、豆角、葫芦，而不是栽培一些争奇斗艳的观赏花，或是营建一个想象的传统园林。

这种现象改变了无界景观对于公共空间、景观设计等诸多问题的看法，颠覆了他们原来推崇的西方现代设计理念。因此他们提出了"隐形的景观设计师"的概念，认识到设计应该去适应设计的对象而不是改变，设计者仅仅是社会生活与经济发展中的协调者。

这次双年展所展出的装置和视频，是无界景观团队夹道花草堂项目的某些意象和前期的结果，展览结束后还将继续。所以在2016年北京国际设计周期间，"胡同花草堂"为杨梅竹斜街的居民举行了"种植展"，以展现普通居民的种植经验和平民的智慧。2017年，"胡同花草堂"在北京国际设计周又为杨梅竹斜街居民举办"'众'瓜得瓜'众'豆得豆"种植展和杨梅竹斜街夹道社区营造项目三年展，并建立了引导居民健身的即时健身系统，进一步鼓励和倡导居民自发的改善居住环境，这些活动颇受社区居民的认同。

当设计师真正关心社会时，就不会只是把自己当成一名设计师，他应当以设计为工具来解决设计以外的事情，因为设计之外的生活更有力量。当今建筑设计、景观设计趋于时尚化、艺术化、奇观化，设计师则像大牌明星，而普通人的生活环境却趋于工业化、边缘化，无界景观工作室的这些探索对中国的设计师来说无疑是一种重要的启迪。他们考虑到人与自然的一种情感上的关联，但不是一种世外桃源的乌托邦梦，因为理想主义太天真了，遥不可及，最平常的东西才是那样亲切、可感。"胡同花草堂"项目以未完成的形态存在，在时间中生长，它重要的功能之一是创造有利于社会交往的空间，重塑胡同中人与人的关系。

由居民共同参加种植活动，在尊重每户人家生活经验的同时营造诗意，共同参与、共享，修复与重建老北京胡同中遗失了的意境。这种平民设计的实践，对北京老城来说，既是雪中送炭，又是锦上添花。

2017 年 9 月于北京

安住 · 杨梅竹斜街改造纪实与背后的思考

The "Hutong Flora Cottage" at the Forefront

Wang Mingxian, Architecture and Art History Scholar
Curator of China Pavilion for the 10th Venice Architecture Biennale

The environmental transformation and continuous community construction of the Yangmeizhu Xiejie has been an experimental project of great importance with regards to the organic renewal of the Beijing old city.

View Unlimited Studio presented a mixed-media installation at the fifteenth Venice International Architecture Biennale, which featured the laneway regenerations that have been applied to courtyards 66-76 in Yangmeizhu Xiejie. The installation was titled "HOME - Communal Garden" and was displayed at the China Pavilliion. The works reflect the urban problems of China from the unique perspective of the dialogue between designers and ordinary people. At the Venice Architecture Biennale, a space that represents one of the most important platforms for international architecture, we were able to have face-to-face communication with members of the international architecture

序

community. The perspectives that they shared were outstanding in their originality, which in turn inspired us with regards to our own forms of thinking.

This year's primary theme at the Venice Architecture Biennale was "Reporting from the Front", which considered six major themes including bionics and the future, war and refugees, informal structures, nature and geopolitics, handicrafts and tradition, home and the sense of belonging. These six themes were utilized to explain the methods of contemporary architecture in dealing with issues such as economic inequality, the uneven distribution of resources, energy consumption, natural disasters and public spaces. The concept of the exhibition being curated at the China Pavilion can be summarized as: "Daily Design, Daily Taoism". This is both a reflection and response to the primary theme of the Biennale, and it is a genuine reflection of the current Chinese social phenomenon and cultural issues present in architecture. The deep social concern that the concept of "HOME - Communal Garden" is based on coincides with the curatorial ideas of the Chinese Pavilion, and echoes the themes present in the academic field. The site construction of the Virgin Garden in Venice reflects an alternative way for Chinese designers and everyday people to embody their own philosophy, and approach towards their work with regards to the status quo of Chinese cities.

The Venice Biennale is the most famous contemporary art exhibition in the world, and is known as the Olympics of the art world. Moreover, the Biennale provides a unique platform for displaying exhibitions. Every exhibition present reflects the frontiers of contemporary art, and indicates emerging trends within the development of contemporary art. The event includes the International Biennial of visual arts and International

安住·杨梅竹斜街改造纪实与背后的思考

Biennial of architecture, which are held alternately. The Venice Biennale was first held in 1980, and in the course of 30 years it has become the worlds most influential event with regards to architectural art and academia. The Venice Biennale is currently the world's top architectural art biennale, and yields an important influence over the development of the direction of global architecture.

In China, large-scale construction can be seen everywhere, and the demolition and transformation has created an atmosphere of utter chaos. "HOME - Communal Garden" considers the development of Chinese cities, and how to allow ordinary people to participate in contemporary architecture. Through the installation of this exhibition, a living exhibition that has continuously grown and changed over the course of the past six months, visitors can participate in planting and harvesting, by which they may experience the interactions that inherently exist between people and design, and the interactions that occur between individuals. The installation is also equipped with live broadcasting equipment that allows participants in Venice to see a live streaming of the Yangmeizhu Xiejie in Beijing, which features a great amount of touching attention to details. Its success has garnered the widespread recognition of the international architecture art community.

As cities increasingly adopt a faster pace of life, Beijing's hutongs have steadily disappeared. It has been no small feat to preserve the remainders of the rare historical artifacts that still exist, such as Yangmeizhu Xiejie alley that is located in the Beijing Dashilan area. The Yangmeizhu Xiejie environmental renovation project began in 2012 and is a multi-level collaborative project. It does not alter the textural qualities of the hutong, and emphasizes elements of modern-day life and modern-day business in conjunction with the traditional lifestyle of the hutong residents. After the implementation of the project, View Unlimited Studio undertook an investigation and discovered that the attitude of the existing residents towards the hutong transformation was reflected in the voluntary behavior of their planting activities. Due to this, there was a significant shift in the central characters with regards to the later stages of the project, focusing on the conception of designs that would facilitate the voluntary creative activity of local residents.

In 2015, View Unlimited Studio began renovations on the courtyards 66-76 of Yangmeizhu Xiejie, with the aim of establishing a "Hutong Flora Cottage", which could act as an intermediary in order to establish an effective way for short-term or long-term residents of the five households to communicate with each other through shared activities such as growing flowers, vegetables, etcetera. In doing so, the aim was to create a community of shared values, so that even temporary residents could enjoy a sense of belonging, whilst also improving the overall quality of living for the residents of the alley. Xie Xiaoying, the chief architect of View Unlimited Studio, stated, Our goal is to upgrade and transform the community, whilst making sure to preserve the original textural qualities of the area. In this way, it may be possible to aid the local residents in moving from a state of "instability", to a state of "abiding", thus establishing an intersection between the declining communities of the old city and contemporary urban life.

The first step was for View Unlimited Studio to implement renovations, including the construction of barrier free facilities and broader passageways, which could improve the overall living environment with regards to common public space. Next, the team needed to establish a flora cottage in the community that could be accessible to all, so that the residents would be able to interact with each other in the form of natural intermediary activities such as growing flowers, planting vegetables, and promoting the healthy development of neighborhood relations. The beautifully designed landscapes and gardening areas that professional designers painstakingly planned out were not of any particular importance to the people living in the bottom strata of society. Some of the residents preferred to plant onions, beans, gourd, and sponge gourd in the planting beds in front of their homes, rather than cultivating the beautiful ornamental flowers that had originally been conceived for the garden.

This phenomenon changed the perspective of View Unlimited in relation to public space, landscape design and many other relevant issues, overturning many of the modern western design concepts that they had previously adopted. Thus, the group put forth the concept of an "invisible landscape designer", and realized that design should adapt to the object of the design rather than enforcing change. Essentially, the designer is simply the coordinator of the site's social life and economic development.

With regards to installations and video featured at this year's Biennale exhibition, the View Unlimited team's work on the alley garden project drew from estimation and early-stage results, which once the exhibition closes, will continue to develop further. Thus, during the 2016 Beijing International Design Week, the "Flora Cottage" held a "planting

序

exhibition" for the residents of Yangmeizhu Xiejie in order to display the planting experience and accumulated wisdom of ordinary residents. In 2017, the "Flora Cottage" was once again featured as part of the Beijing Design Week, during which a planting exhibition titled "Together we sow, Together we harvest" was held on behalf of the residents in Yangmeizhu Xiejie, which integrated an exhibition displaying the three-year period of renovations. These renovations include a comprehensive fitness facility that is yet another step towards encouraging locals to actively improve their living environment. Activities such as these are these are well recognized by the community.

When a designer really cares about their society, they will not view their role solely through such a narrow lens. Rather, they ideally should use design as a tool to solve issues that are beyond the narrow scope design, because there is a life beyond design that is even more powerful. Today, architectural design and landscape design is moving further towards the realm of fashion, art, and spectacle. Designers are like giant celebrities, however the lives of ordinary people and their living environment have become increasingly industrialized and marginalized. View Unlimited Studio's decision to explore this aspect will undoubtedly hold the potential to aspire this generation of Chinese designers. They consider the emotional connection that exists between humans and nature. However, it is not a paradise that they dream of achieving, because such idealism is too rich in naivety, and such a goal is unreachable. Rather, the most common quality is to be kind and sensible. The "Flora Cottage" project exists in an uncompleted form and will continue to mature over time. One of its most important functions is to create space that is conducive to social interaction and to reshape the relationships between people residing in the hutongs. The residents participate in planting activities together, respecting the experience of each and every family, and in doing so, are creating living poetry together by jointly participating and sharing together in the restoration of the old Beijing alleys. For the city of Beijing, this practice of civilian design in action is both a timely assistance, and also the metaphorical icing on the cake.

September 2017, Beijing

序二　安住

——无界景观设计团队参与的北京老城区有机更新

赵园

著名文史研究者

较之无界景观工作室已有的设计作品，他们承接的北京大栅栏片区杨梅竹斜街改造项目，更具公益性质。该项目是由市政当局主导，设计团队在完成有关任务后，继续延伸，将他们的思路实现在街巷深处居民"生境"的改善上，使团队一向坚持的理念进一步落地，以此探索老城区改造的路径，也作为回馈社会的一次努力。项目后续延伸的部分，入选了2016年威尼斯双年展。正是这种与基层民众的亲和性、设计风格的平民性，使项目具有了感染力，与双年展的宗旨有不谋之合。

杨梅竹斜街改造及后续投入，关系庶民的日常生活。在中高收入者陆续迁往新兴社区后，留在胡同中的居民，多为偏低收入者、老人及租户。设计团队的关切是在居住条件尚未根本改善的条件下，使留在胡同的居民得以安住；以较低的成本，为由于快速城市化、商业化造成的生活质量、生存境况的落差，提供（即使是有限的）补偿；在公共投入不足的情况下，整合、梳理有限的资源，使居民由优化居住环境受益，而非一味将"美好生活"期之于未来。

在街巷改造中，设计团队为自己设定的任务是"厘清并保护胡同街区中固有的文化基因，并通过设计弥补其基因缺陷"。在不改变胡同肌理的前提下，以新旧并置，使时尚元素、新商业业态与居民原有的生活和谐共生，一并成为居民日常生活的构成部分。除保存原有地标性历史文化建筑、推动已有商业设施转型升级外，团队将设计工作落在改善居民日用设施、铺设渗水透气的地砖等处。以"微创与介入式为胡同中的住户提供针对性的设计"，逐户整修老人日常休憩的门前台阶，以沿街花坛为街道增绿，着力于"新材料与历史材料的编织"，工程的每一局部均不吝细心打磨。即如取特殊烧制的地砖而非水泥、沥青，且以艺术化的铺设传递文化信息，使人们日日行走其上的路面成为怀旧的诱因。能呼吸的街道有助于小气候的改善。居民说新铺的地砖暑天不热蒸，即对此功能性层面的体验。这种体验自然由比较中来。所有这类看似细微处，非由在地居民的角度则不能体会。

回头看，我居住的"老旧小区"上一轮的"改造"中，倘也能交由有创意的景观设计团队，逐户征询居民意愿，因楼施策，而非使用统一方案，结果会不会有所不同？我曾在其他处谈到东邻日本对人居环境的精心营造——绝不因空间狭小即粗率应付，也因此随处可见小巧精致、各具面目、赏心悦目的私人庭院与公共空间。这是一个有高度审美修养的民族对于日常生活的态度。我个人的审美偏好，在大山大水，对江南园林几近无感。但改变城市的狭小空间，却有必要借鉴古代中国的园林艺术，即使只能取其缩微形态。如果我们的景观设计师能如古代文人打造自家园林那样塑造城市，城市的样貌当有怎样的不同！

在这个设计团队看来，"宜居城市"应当使居住其中者都能感受到"环境友好"。老北京曾掩映在绿树浓荫中。普通人家、寻常院落，往往都有一棵大槐树或老榆树，在自家门口即不难亲近自然。

设计团队完成了街道改造的一期工程后，二期工程中更向公益的方向延伸，主动承担了部分"志愿者"的角色，深入住户的庭院、夹道、隙地，以不期待回报的公益精神组织居民改善其生活环境，从中培育公共精神；以润物细无声式的渗透、浸润，经由点点滴滴的改变，使共享、分享的概念深入人心。

北京老城区街巷房舍密集，公共空间狭窄细碎，宜于逐点逐线展开设计构想，整合散碎空间；以居民间的相互协调、自主参与，见缝插针地栽培种植，使居民由自家门口的花木感受生活之美；经由共同种植拉近相互间的距离，营造如老北京胡同曾经有过的和谐的邻里关系；在远离山水田园的地方，以绿色空间获取新的生存体验。在这过程中，团队极其强调的是居民自主自发的行动中产生的参与感。这是生成"家园"意识、培育对街区归属感的关键条件，也是激发老城区活力的关键条件。在项目实施的过程中，让设计团队印象深刻的是胡同中的人与人的关系，胡同居民朴素（甚至卑微）的对生活的期待。这里仍然是有故事的胡同，不断衍生着新的故事，如老舍笔下的胡同那样。上文说到设计团队不期待回报，但是在改造过程中，居民作为社区主体的意识逐渐增强，对于改善自身生存条件充满向往。倘若上述意识与向往持续传递，岂不是对设计团队最好的回报？街道改造项目的实施，基于市政当局的财政支持与相关机构的合作；后续的部分，更系于与在地居民的沟通与互动。居民成为设计的参与者与实施者。对普通居民人伦日用的体察，身临其境、设身处地地与居民沟通——对于设计团队，无疑是新的经验。

威尼斯双年展的参展项目所依托的，正是杨梅竹斜街的一段夹道，五户人家共用的狭小空间。在这样的空间营造诗意、改善居民"生境"的,是居民共同参与的种植活动,其结果——即参展作品——甚至不是"最终的",因其需要持续地投入、经营,在时间中"生长"。该项目的创意也正在于此。共同参与、共享,与其说是物质意义上的,不如说更是文化层面的——修复与重建老北京胡同文化中遗失了的意境。设计团队属意的是当地居民的获得感:即使依旧陋巷陋室,却能在此安住;在物质条件仍然简陋的情况下,拥有一份有尊严的生活,并对继续改善抱有期待。

双年展后,项目在公益的方向上延续,即被设计团队称作"平民花园"的"花草堂"。设计团队说他们参与建造的"平民花园","没有成本,没有设计","没有刻意的视觉营造","无需环保主义的说教与劝导"。他们珍视的是"平民日常生活的美学",是平民的那种"看似与发展主义逻辑相悖的日常生活实践",相信这种实践"能够为迷茫于当下消费社会中的人们提供另外一个参照系"。而那些散布在胡同居民房前屋后隙地上的,的确有十足的"平民性"。不招摇,不张扬,细碎而平常。出门即见,可随时驻足欣赏,与邻居小聚。点状片状的花木,使逼仄隘陋的居民区随处绿意充盈,生机盎然。设计团队所期待的是点滴的改善,经由点滴改善达成的累积效应。他们的目标更在营造和谐的有利于社会交往空间,重塑胡同中人与人的关系;在协同改善胡同生态的过程中,凝聚而成居民的共同体。设计团队自身也由上述过程中获益，不但收获了设计灵感，而且收获了公益精神。在与居民沟通、互动中形成"设计"，是新鲜的经验；作品以未完成的形态存在并期待其继续发展，也是新鲜的设计思路。中国的文化传统注重"人伦日用"。设计团队的理念正与此契合。"老北京"的风味不止要有胡同、四合院（大多早已沦为大杂院），更要以胡同居民的生存状态、邻里关系、生活氛围等体现。否则徒有其表，"老北京"的气韵何在！

无界景观设计工作室既不求速效，也相信不会速朽。生活中任何微小而切实的改变，对于底层民众都意义重大。景观设计本来就无须一味追求煊赫，其目标或更应设在以其工作提升人的生活质量，改善人的生存状态，使生活在城市角隅的人们也能享受现代文明的成果。设计者应当不远于"人间烟火气"，使其设计思路贴地、贴近居民生活，尤其是基层民众的生活。这里的启示或许在于，即使有资质、有能力从事大项目的设计师、设计团队也不妨放下身段，在自己居住的城市施展拳脚，将设计理念实现在普通居民的生活空间中。

"安住"，是升斗小民朴素至极的向往，几难称"理念"。但"安住"何尝容易！是否"安"，端在居住者的感受。豪宅并非就可以安住，陋室也非不可安住。陶诗所谓"暖暖远人村，依依墟里烟"（《归园田居》），"田夫荷锄至，相见语依依"（《渭川田家》），是谓安住。留在日见破败的胡同中的居民能否"安住"，是市政当局的一大难题。使低收入者以及临时落脚的打工者也能"安住"，属于市政当局所应关怀的最基本的民生，基础性建设，也是实实在在的"政绩"。

作为人文学者，我的专业与这个团队的设计活动并无交集。如果一定要找一个交接点，那就是完稿于20世纪80年代末、出版于20世纪90年代初的那本《北京：城与人》。长期以来，我只是从旁留意他们的活动，多少为自己打开了一个关注与思考的空间。而大栅栏片区杨梅竹斜街改造这一项目，使我在《北京：城与人》后又回到了疏离已久的老北京胡同这一主题，那个场域借此重回我的视野，有久违重逢的亲切感。我自然希望这个团队的理念被更多的景观设计从业人员与关心城市建设的人们关注，不断衍生新的公益项目与文化产品，让老北京的故事得以延续，改善老城区居民生存状况成为吸引更多人致力的事业。

老北京也如中国的其他老城，不同社会阶层混居。1949年后，老北京的世俗性格、平民气质仍有延续，一段时间里保持了混居状态，尽管其中有了"大院"，大宅门则成为官员的宅邸。20世纪90年代以降，少数胡同进行了适应旅游业的商业化改造，其他胡同则基础设施老旧，在日甚一日地衰败。传统街区从来是这座城市主体的构成部分，却大部分败落为背街小巷，成为商业建筑的黯淡背景。近几十年来高档住宅区的兴起，固然势所必至，其社会成本却没有引起足够的关注，即如将社会分层体现于城市的空间分割、分隔。这种情况或方便了市政当局缓解老城区改造的压力，却也将老城居民生活状况的改善无限期地后延。

对于无界景观设计团队参与的老城区改造，我欣赏的更是"有机更新""基因修复"的思路。有机更新，即在不破坏胡同原有肌理的前提下更新；基因修复更着意于老北京历史文化的存续。专业性、技术性与人文性并重，将人文融入专业设计，这一思路值得继续阐发，被后续的老城区改造吸纳。却也应当说，设计团队的工作带有一定的实验性；设计团队对五户人家种植活动的介入也有样品性质，其推广尚需更多公益组织与志愿者的参与；经由共同种植建立的平台亦不免脆弱，随着人员的流动与邻里关系的变化，其维系也需要相应的条件。北京老城区是大社会，其中有诸种小社会，差异丰富。在深入具体街区的基础上，逐街逐巷逐院逐户地调研，针对性地设计，工作量之大，非一个设计团队所能承担。老城区的更新改造起步未久，后续工作烦难而琐细，是不折不扣的浩大工程。

此外还应当说，公益项目不能替代服务设施的优化升级。后者是市政当局不可推卸的责任。将社会平等落到实处，需要对低收入者聚居区、外来务工人员聚居区加大财政投入，像在农村地区扶贫攻坚那样。相对于这种更新改造，大拆大建毋宁说更容易，也为地方政府与房地产商所热衷。老城区的改变见效慢，少了"能见度"，难以用光鲜亮丽吸引眼球。其迟迟不见推进，原因端在于此。上述这些，属于我个人的延伸思考。我虽不居住在胡同，却住在始建于20世纪80年代上半期的"老旧小区"，较之高档社区的住户，或更能体会"老城区"居民的处境。社会分化不可避免。将级差控制在一定范围，至少有利于社会稳定吧！

如上所说，北京胡同居民的成分早已在市场化的过程中改变。此胡同已不是彼胡同。如何为现有的胡同保存生机，是一个有待持续关注的课题。北京不应当如某些城市那样，仅仅将尚存的街巷作为镶嵌在现代城市中供游客玩赏的"老街"，而是在现有的胡同中保存曾经有过的丰厚的文化意蕴，才是真正的挑战。

"安住""生境"，都属于无界景观设计工作室为自己所拟核心概念。为安住而设计，为营造、改善人的生境而设计，为更多人共享发展成果而设计。一个城市不应当有被"发展"遗忘的角落，北京尤其不应当有。北京胡同的世俗性格，胡同文化的平民品质，不应当被拔地而起的政府机关大楼与商业大厦覆盖。胡同是人境，无论是原住户还是外来者，房主还是租户，以及流动不居的务工者，都有理由保有对合理生活的期待。无界景观设计团队相信，经过"有机更新""基因修复"的胡同，依然有可能恢复并体现这座城市曾经有过的特质。

近闻北京城市总规划将"旧城"改为"老城"，以体现对城市历史积淀的尊重。以"老"代"旧"中，有观念的调整：由弃旧如敝屣，到尊重、保存历史文化遗留。在一轮轮大拆大建之后，终于迎来了转机：关于城市改造，关于"老北京"，关于"老城区"。尽管来得太迟，总比固守已有的"发展"思路要好。"尊重"之余，维护、升级改造的任务提上日程。无界设计团队参与的杨梅竹斜街改造，或能提供某种参考。

（本文中的引文引自无界景观设计室项目说明）

2017 年 7 月于北京

HOME – View Unlimited Landscape Design Team's Reconstruction of the Beijing Old City Area

Zhao Yuan, renowned literature and history researcher

In comparison to the existing design works of View Unlimited Landscape Architects Studio, with regards to the undertaking of the Yangmeizhu Xiejie renovation project in the Dashilan area, the project possessed a greater public interest aspect. The project itself was dominated by municipal authorities, thus after the completion of the design team's work, additional aspects remained that were an extension of the project – namely the conceptual ideas which were realized in the improvement of residents' local "habitat", the streets and alleys of their everyday life. This inspired the team to consistently adhere to the concept of greater localization in order to explore the most ideal methods of transformation of the old city, in addition to making a concerted ef-fort to give back to society. In addition, the subsequent project was selected for an award at the 2016 Venice Biennale. It is this affinity with the grassroots people and everyday design aesthetics that created a contagious atmosphere among all participants involved with the design, and which was directly in alignment with the principles of the biennale.

The transformation and follow-up involvement of Yangmeizhu Xiejie is directly related to the residents' daily life. After a steady progression of middle and high-income earners moving to new communities, the present hutong residents are generally comprised of low-income earners, the elderly and outside renters. The primary concerns of the design team were that living conditions have failed to be improved consistently over time, thus being unable to facilitate a high level of safety with regards to the living standards of local residents. By utilizing minimal costs, in order to spur the rapid onset of urbanization, the effects of commercialization have led to a drastic reduction in the quality of resident's living conditions. This has resulted in (limited) compensation; under the conditions of relatively insufficient public investment, the integration and management of these limited resources has allowed for residents to benefit from these more optimal conditions rather than placing all of their hopes and dreams towards the future prospects of "a glorious life ahead".

During the process of the street and alley renovations, the design team conceived of a personal set of goals which were to be pursued; namely, "to clarify and protect the inherent cultural imprint that is present in the hutongs, and apply new designs in order to makeup for any present defects." Without changing the overall aesthetic of the existing hutong fabric in any noticeable way, the juxtaposition of the new and old, modern fashionable elements, and the harmonious existence between new forms of businesses and residents' original lifestyles, has resulted in the components of everyday life that surround residents. In addition to preserving the original landmark historical and cultural buildings and promoting the transformation and upgrading of existing commercial facilities, the team aimed to conceive of designs that would improve the everyday facilities of residents, such as paved tiles that would be more effective in drainage, etc. By "utilizing minimally invasive and minimally intrusive methods to provide targeted designs to the hutong households", restoring the front steps outside of each household entrance which are preferred by the elderly as a resting place, installing flower beds along each side of the alley for the means of beautification, and focusing on the "interweaving of historic and modern elements", all of the intricate details of the project were lavished with careful attention. For example, the paved tiles were created by a special firing process rather than by using cement or asphalt. By utilizing this special process with regards to the installation of these tiles, it was possible to convey an inherent cultural knowledge. In doing so, the road that people walk along day-by-day may adopt a nostalgic quality. By applying technology that may allow the paving to be more breathable, the positive effects of this technology may influence the microenvironment. Residents have reported that the tiles are not steaming hot like those in summers past, thus showing that such designs are part of a multi-tiered functional experience. Moreover, such experiences are naturally based on comparisons with past interactions. Thus, it is the local inhabitants who primarily experience all of these seemingly subtle additions.

Looking back on the previous round of "transformations" which have been applied to my old "stomping grounds", it is fair to consider if it were possible to hand the project to a creative design team, which by taking into consideration the wishes of the residents, rather than applying a unified methodology because of building policies, would the results be any different? Previously, I have talked about the elaborate creation of human settlements in our neighborhood country - Japan – despite the limited space available, this had no bearing on their ability to deal with the issues associated with elegant development. As a result, diminutive and exquisitely designed areas can be found everywhere, including both private courtyards and public spaces. This is essentially the result of a highly aesthetically oriented nation's attitude towards the facets of everyday life. Because of my personal aesthetic preference, that of far reaching mountains and natural water features, I have very limited feelings for the features of the traditional Chinese gardens in the south. However, if there were a means to alter the limited spaces in our cities, it is necessary to draw lessons from ancient Chinese garden art in order to take advantage of limited spaces. If our landscape architects are able to shape cities like the ancient literati built their own gardens, how different would our modern cities differ in appearance!

From the view of the design team, the principle requirement of "livable cities" is that the residents feel that their surrounding environment is of a "friendly nature". The Old Beijing of years gone by was once hidden in the shade of trees and greenery. The average family would often feature a large locust tree or an old elm in their simple courtyard space. There are no barriers in getting close to nature when it is at your own doorstep.

After the design team completed the first phase of the street renovation project, the second phase of the project extended into the realm of public welfare. Essentially, the team voluntarily took on the role of "volunteer"; entering into the courtyards of households, narrow lanes and open spaces, bringing with them a selfless spirit of public service. The team organized residents in order to improve their living environment, in doing so cultivating a public spirit. By organically embedding themselves among the residents, and little by little applying gradual changes by a process of sharing and discussing opinions, the team was able to genuinely connect with the residents.

In the old city area of Beijing, the streets and houses are densely packed; public space is at a premium and extensively fragmented, thus it was necessary to conceive of a design that may be specifically implemented in order to comprehensively integrate a scattered space. By coordinating residents who participated independently, and by making use of every modicum of available space for cultivation, the residents were able to appreciate the beauty of their everyday life by means of their own flowers and plants situated at the entrance of their residences. By creating a public area in which to cultivate greenery, it was possible to narrow the distance between residents and create a harmonious neighborhood similar to that which existed in the old Beijing hutongs. Far away from the vast, pastoral areas, the designers successfully utilized greenery in order to cultivate a new type of everyday existence. During the process, the team placed a great emphasis on the involvement and spontaneous actions of the residents, in order to foster a sense of participation and action. By doing so, the team cultivated a concept of "home" – a key condition for both fostering a sense of belonging within the neighborhood, and stimulating the vitality of the old city. During the process of implementing the project, the design team was left with a strong impression with regards to the inter-personal relationships between hutong residents and their simple (or even humble) expectations towards life. This is still a hutong that is bursting with stories, ever continuing to derive new stories similar to the alleys featured in the works of Lao She. As previously mentioned, the team did not expect any "returns" on their hard work; it should be said that any "returns" is simply the consciousness of the residents with regards to assuming the main body of the community and subsequently fostering a yearning for the improvement of their respective living conditions. If it is possible for this new consciousness and yearning to take root among the community, is the not the most valuable reward possible for the design team? With regards to the implementation of the renovations project, the formative stage was based on the financial support of the municipal authorities and cooperation with relevant agencies. The subsequent stages however, shall be more closely related to the communication and interaction of those involved with local residents. In essence, residents shall become the participants and implementers of the design. As for the daily experiences of ordinary life among local residents, the design team's efforts to place themselves in the shoes of their counterparts were unquestionably an entirely new experience for all those involved.

The exhibition featured at the Venice Biennale is based on one section of Yangmeizhu Xiejie, a narrow space shared by five families. Creating such a poetic atmosphere, and improving the residents' habitat was conceived of as an activity that residents could participate in jointly with the design team. The result – the exhibited works-are not a "final result", because the project itself requires a continuation of investment, management in order to "grow" over time. The core essence of the project may be found in this. Joint participation and sharing – not in a non-materialistic sense but rather in a cultural sense by means of the restoration and reconstruction of lost artistic concepts that are native to the in the old hutong culture of Beijing. The design team's principal desire was to provide a sense of gain to the local residents; although the environment may still remain in essence an alley, it may be livable. Under the condition that the material conditions are still elementary, it may be possible to live a dignified life. With regards to this, the team looks forward to a sustained improvement that may be gradually carried out.

After the Biennale, the project has continued in the direction of public welfare, having been termed "Communal Garden" by the design team. In participating in the creation of this "garden", the design team have stated that "invisible design that cost nothing". Moreover, "without a deliberate visual creation, there is no need for any persuasive ecological moralism". The design team's most cherished aspect is the "aesthetic understanding" that has been adopted in the daily life of the residents, namely practicing a style of daily life that is contrary to the "doctrinal logic" of development. The belief is that this type of practice can provide another frame of reference for people who are lost in the current consumer society. As for those open spaces that are scattered around the front and back of hutong residences, indeed they boast a particular form and quality of the "common citizen". Each minute section is unassuming and unobtrusive in appearance, and entirely ordinary. One need only to go outside and take a moment to appreciate the scene of local neighbors socializing together. The varied greenery creates the effect of a flourishing natural environment, full of vitality, within the confines of a cramped and crowded area. The chief desire of the design team is to witness gradual improvement, which little by little may achieve a cumula-tive effect. Their goal is to create a harmonious society that is conducive to social interaction and which may reshape the interpersonal relationships between residents of the hutongs. Throughout the process of collaboratively developing and improving the overall ecology of the hutongs, the residents transformed into a cohesive community. The design team itself also benefited from the process; not only gaining inspiration from the designs that have been conceived, but also from the spirit of public service. Whether it be communicating with the local residents, conceiving of "designs" through interaction has been a refreshing and invigorating experience. At present the works exist in an unfinished form, and it is expected to continue developing. This in itself is also a fresh and inspiring design concept. China's cultural tradition attaches significant importance to "human relations and daily use (The Daily Taoism)". The concepts applied by the design team are consistent with this. The concept of "old Beijing" is not limited to the aesthetic appearance of the hutongs and courtyards (of which most have been altered into tenements), but more importantly includes the inherent standard of living, relationships that exist between neighbors, the overall living atmosphere, etc. If the quality of the hutongs was contingent on aesthetic appearances only, what type of "old Beijing spirit" would remain!

View Unlimited Landscape Architects Studio did not seek out quick solutions for the project and believes that because of this it will stand the test of time. Any tiny and tangible change in one's life is of great importance with regards to those people who will be affected at the grassroots level. It is not necessary for landscape design to pursue a grandiose style; rather, the core goal of any project should be focused on improving the quality of people's lives and their overall living conditions. The people living tucked away in the corner of the city can also enjoy the achievements of modern civilization. Designers should never distance themselves from the "earthly life" of everyday. Rather their design principles should be aimed towards the grassroots level, directed as close as possible to the everyday life of local residents, especially with regards to the most basic levels of people's lives. The inspiration here may lie within the fact that even qualified designers and design teams who have the ability to undertake large projects, may drop their head in despair at the scope of difficulty with regards to such a project. Nevertheless, when it is one's own home, the only option is to raise one's fist and apply a design concept that is realized in the living space of ordinary residents.

To "abide in", feeling a sense of "home" is a strong yearning which is pos-sessed by ordinary citizens, and could even be considered a "conceptual idea". However, to "abide in" is no easy feat! Perhaps the concept of "abiding in" may only be interpreted in a subjective manner by the resident themselves. Whereas a mansion may perhaps not necessarily feel safe to reside in, this concept is no different in its application to a shack or any other form of residence. The so-called "warm village far away, ever emitting smoke" ("Return to Nature") and   the "farmers with his hoe, who upon meeting naturally picks up his conversation" mentioned in Tao Yuanming's poetry, and are two such examples. In the case of the increasingly dilapidated hutongs, is it possible for the residents to have a sense of "abiding in"? This is indeed a serious issue to consider for the municipal authorities. Considering that low-income migrant workers tend to take up temporary residence in these areas in order to also have a place to "abide in", the city should be concerned about the most basic livelihood of these people. The development of basic construction infrastructure is the most obvious way in which to base any degree of achievement with regards to this issue.

As a humanist, there is very little intersection between my profession and the activities of the landscape design team. If I were to have to choose one work, it would have to be "Beijing: City and People", which was written in the 1980's and finally published during the 1990's. For a long time, I simply paid attention to their activities as an observer, at the very least developing a certain invested interest. Moreover, with regards to the renovation project of Yangmeizhu Xiejie in Dashilan, I felt that after encountering "Beijing: City and People", I was once again facing the theme of being estranged from the old hutongs, a scene that I was only now once again being confronted with. The feeling may be described as one of intimacy after a long absence. Naturally, I hope that the concepts held by this design team, and their concern for urban construction may be adopted by a greater number of landscape architects. Furthermore, the hope is that these design teams may continue to constantly derive new public welfare projects and develop cultural products, in order to allow the story of old Beijing to continue. In doing so, it may be possible to attract a greater number of participants who may be involve in improving the living conditions of the old city residents.

Old Beijing is rather similar to other old sections of cities in China; of course there are ghetto-like area (such as the north part of the city), in

序

addition to a variety of hutongs which embody the varied social classes that reside within. After 1949, the secular character of old Beijing and the steadfast temperament of its residents has been maintained as a mixed sort of state. Although there are "courtyards", the grand mansion gates have officially become a signature element of the abodes of officials. After 1990, some of the hutongs have adopted commercial reforms in order to develop a tourism industry. As for the other hutongs that have failed to adopt these reforms, the infrastructure remains old and dated and is increasingly growing more and more decrepit. The grid-like set of streets has been a traditional hallmark of the city, however behind this giant grid are the arteries of the small alleys and hutongs, which function as a dim backdrop to the commercial activities of the city. The rise of high-income residential areas in recent decades is bound to lead to certain developments, however the social costs have failed to attract sufficient attention. For example, the social stratification of the city has become reflected in the division of certain sections within the city. This situation has facilitated a lack of motivation among the municipal authorities with regards to carrying out any necessary renovations of the old city. However, under such conditions, as a result the improvement of the living conditions of the old city residents has been postponed indefinitely.

As for the participation of View Unlimited Design Team in the renovations of the city, I appreciate the concept of "organic restoration" and "gene recovery" which they have applied. Organic restoration is related to performing restorations without changing the original qualities of the Hutong, whilst gene recovery focuses more on the history and culture inherently present in the existence of old Beijing. By placing equal emphasis on specialist skills, technology and humanism, it has been possible to integrate humanism into professional designs. This concept deserves further elucidation, in order to be further applied within the continued renovation activities in the old urban areas. However, it should be said that the design team works with a certain degree of experimentation: the design team also has sample properties for the intervention of five households' planting activities, at present the activity requires more public welfare organizations and volunteers to participate in its promotion; platforms for municipal greening cannot help but be vulnerable, due to the constant flow of new people and the changes between neighborhood relations, its maintenance requires a corresponding set of conditions. The old areas of Beijing city are comprised of a large society, of which a rich and diverse set of smaller societies exist within. Simply on the basis of a specific set of street blocks, which includes each and every secondary street, alley, courtyard and household, to conceive of designs whilst keeping all of these elements in mind is a huge undertaking – one which a design team is unable to solely bear the responsibility for. Not long after the old city renovations first began, the follow up work became increasingly complex and trivial in nature. Indeed, such work reflects the enormous scope of a project of this nature.

In addition, it should also be said that public welfare projects cannot replace the optimization and upgrading of service facilities. The latter is an inescapable responsibility of the municipal authorities. Putting social equity into practice requires increasing the financial input directed towards the areas inhabited by low-income people and migrant workers, similar to poverty alleviation in rural areas. With regards to these renovations, a large demolition is far more easy to undertake, and is an option that the local government bodies and real estate developers are keen on. Changes in the old town are generally slow in nature, with less "visibility". Because of this, it has not been suitable to apply bright and eye-catching methods that attract attention. Subsequently, this is a key reason as relating to the delays that have occurred with regards to improving the area. The aspects mentioned above are simply an extension of my own thinking. Although I do not live in the hutongs, I live in an "old city" which was built in the first half of the 1980s, and the residents in the "old city" are better able to understand the situation of the "old city" rather than those living in upscale communities. Social differentiation is inevitable, however containing the level of stratification within a certain range may at least be conducive to social stability.

Based on the aforementioned commentary, through the process of marketization the composition of the residents within the Beijing hutongs has already experienced great changes. This alley is no longer the hutong of yore. The question of how to preserve the way of life in the existing hutongs is an ongoing one of vital importance. Beijing need not model itself on other cities, the only remaining original streets and alleys being transformed into tourist sites, called "old streets". Rather, the true chal-lenge is to preserve the rich cultural connotations that presently exist in the hutongs.

The concepts of "Abiding in" and "habitat" are both considered core concepts by View Unlimited Landscape Architects Studio. The team has conceived of designs with an eye towards livability, which through their construction may improve the lives of residents, and by which a greater number of participants may share in the fruits of development. Regardless of "development", a city should allow any part of itself to be forgotten. This is especially the case as relating to Beijing. The secular character of hutongs in Beijing and the unique qualities of the inhabitants that hutong culture is comprised of should not be covered by municipal government and commercial properties. The hutongs are defined by the inhabitants which they are populated by. Whether original inhabitants or outsiders that have taken up residence, homeowners or tenants, or migrant workers who have unstable living circumstances, there is no reason for any of these actors to fail to maintain their hope for improvements in their living environment. View Unlimited Landscape Design believe that by utilizing the concepts of "organic restoration" and "gene recovery" with regards to the hutongs, it is possible to restore and once again embody the qualities that the city once had.

Recently, the Beijing city administration has decided to rebrand the "old city" with a syntactically similar moniker in order to reflect and confer honor on the historical value of the city. By replacing the term "old" with the more venerable phrase of "long standing", there has been a conceptual adjustment: by discarding the old and dilapidated, it may be possible to honor and preserve the historical and cultural legacy of the area. After a round of major demolitions and construction, a turning point has finally arrived: with regards to urban transformation, "old Beijing" and the "old city". Although late in its arrival, it's better than sticking to the pervious idea of "development". The path to "respectability", maintenance, and renovation upgrades has found its place on the agenda. Is it possible that by participating in the Yangmeizhu Xiejie renovations, View Unlimited Design Team may provide a point of reference moving forward?

(The citations in this article are illustrated in the project featured by View Unlimited Landscape Architects Studio)

July 2017, Beijing

目 录

第一部分　方案实施　/ 001
一　概述　/ 003
二　杨梅竹斜街有机更新（2012—2013）　/ 046
三　胡同花草堂（2015.7—2015.9）——杨梅竹斜街 66—76 号夹道社区营造计划　/ 074
四　杨梅竹斜街社区治理与公共空间提升（2017—2018）　/ 090
五　一尺大街改造设计及实施情况（2012—）　/ 106

第二部分　展览传播　/ 117
一　安住·平民花园——2016 年第 15 届威尼斯国际建筑双年展中国馆作品　/ 119
二　杨梅竹花草堂 2016——日常生活的景观　/ 217
三　杨梅竹花草堂 2017——"众"瓜得瓜"众"豆得豆　/ 230
四　三岁，胡同花草堂——杨梅竹斜街 66—76 号夹道社区营造项目三年展　/ 242
五　软组织·胡同中的即时健身系统　/ 246
六　杨梅竹斜街小气候监测数据交互展　/ 250
七　杨梅竹花草堂 2018——繁华落尽终归平常　/ 253
八　"花草堂"的延伸　/ 259

结语　/ 275
2012—2024 年杨梅竹斜街环境更新大事记　/ 290
项目团队　/ 297
无界景观设计工作室　/ 299
致谢　/ 300

Contents

Part 1 Implementation   / 001
01 Summary   / 027
02 Yangmeizhu Xiejie's Organic Renewal（2012-2013）   / 046
03 Hutong Flora Cottage（2015.7-2015.9）--Public Space Renovation of 66-76 Yangmeizhu Xiejie Alley   / 074
04 Yangmeizhu Xiejie Community Governance and Public Space Enhancement（2017-2018）   / 090
05 Renovation and Transformation of Yichi Street（2012-）   / 106

Part 2 Exhibition Communication   / 117
01 HOME Communal Garden--The 15th Venice International Architecture Biennale China Pavilion,2016   / 119
02 Yangmeizhu Xiejie Hutong Flora Cottage 2016--Daily Tao VS Spectacle   / 217
03 Yangmeizhu Xiejie Hutong Flora Cottage 2017--Together we sow, Together we harvest   / 230
04 Three-Year-Old，Hutong Flora Cottage--Three Years of Revitalizing the Alley 66-76 Yangmeizhu Xiejie   / 242
05 Soft Tissues: Spontaneous Exercise System in Hutongs   / 246
06 Exhibition: Interactive Display of Microclimate Monitoring Data on Yangmeizhu Xiejie   / 250
07 Yangmeizhu Xiejie Hutong Flora Cottage 2018--When All Comes Down to Ordinary   / 253
08 Flora Cottage Expansion   / 259

Conclusion   / 282
Highlights from 2012 to 2024: Environmental Renewal of Yangmeizhu Xiejie / 290
Project Team   / 297
View Unlimited Landscape Architects Studio, CUCD, CCTC   / 299
Acknowledgments   / 300

"我从小在这里长大,现在这条街变成'文化街'了,和我没有关系了。"——杨梅竹斜街一居民(2015)
I grew up here, but now this street is totally unrelated to me. It has become a "culture-site". —said a local resident at Yangmeizhu Xiejie(2015)

第一部分
方案实施

Part 1 Implementation

第一部分　方案实施

一、概述

杨梅竹斜街改造项目是北京老城改造工程的一部分，是一个由地方政府牵头、以开发商的商业模式运作的项目。项目的预期目标是：在不破坏原有老城街道肌理的基础上，改造街道的基础设施与环境外观，通过房屋置换等手段引进新的商业机制，将这条以本地居民为主的老街打造成具有自我生产能力的旅游产品。从项目策划到项目实施，参与其中的政府部门、开发商、建筑设计师、景观设计师、街道居民始终处于基于各自目标、利益、趣味的博弈之中。从项目的完成效果看，它无疑是各方相互妥协的结果。

与紧邻的大栅栏西街相比，改造完成后的杨梅竹斜街显得清净许多。这里既没有各种地方特色的店铺、餐馆和旅游纪念品商店，也少有不同口音或不同肤色的游客。除一年几次的诸如"北京设计周"等公共文化活动会给这条街道

安住·杨梅竹斜街改造纪实与背后的思考

举办"北京设计周"时的喧闹和平日略显冷清的杨梅竹斜街
Yangmeizhu Xiejie during the hustle and bustle of "Beijing Design Week", and appearing relatively deserted on a typically peaceful day

带来短暂的喧闹外,这里的日常生活呈现着平淡的景象,居民的街谈巷议也大多操着老北京南城口音。

从改造完成后的街道总体风貌看,除了贯穿整个街道的约四百多米长的特殊设计的地面铺装、台阶和标志有"大栅栏"字样的井盖,以及几家特征明显的时尚小店、咖啡馆、酒吧、书店外,似乎也看不出更多被设计的痕迹,其中某些地段甚至仍然略显杂乱。从街道绿化的角度看,除了每年举办的北京设计周及在重大节假日期间市政园林部门摆放的盆栽植物和花草外,平日里的杨梅竹斜街显得有些"荒凉"。在设计师出于绿化和美化街道环境目的而精心设计的屋边花池中,有些植物已面目全非,部分花池甚至被居民们改种成自家的丝瓜、小葱等。从设计的角度看,改造完成后的街道与国内各大城市中的一些著名"老街巷"或"艺术区"相比实在不可同日而语,甚至找不到几处能够进入摄影画面的视角。然而,这短短 500 米的地面铺装与局部的房屋改造,从规划、设计到施工完成却用了近 3 年,其中大量的工作都是用在协调街道设计的总体风格与街道两侧居民的利益与趣味上。

如前所述,杨梅竹斜街改造是牵涉各方的意图、目标、利益之间的冲突与妥协之下的成果。其中最值得关注与反思的是本地居民在这一公共项目中的态度。按照委托方的要求，设计方在街道的地面铺装设计中采用了含有传统工艺和模数的耐压地砖，其目的是与一些仍保持原状的老建筑衔接并能承载汽车的压力。由于街道原有的绿化环境极差，几乎没有乔木类植物，因此，在绿化设计中增加了许多与居民房屋衔接的种植灌木植物的花池以提高街道的绿化率。但是所有这些出于美化环境目的的举措并没有受到多数居民的支持。根据施工记录，施工完成后，花池内种植的植物需要经常更换，原因是这些植物经常性地被居民采掘或破坏，部分花池中甚至被居民栽种了自家的实用类植物如小葱、丝瓜、豆角等。除此之外，根据市政绿化单位的报告，每年举办各种公共活动期间摆放的盆栽植物有20%被当地居民搬回家。

以上这些负面信息，如果站在公共媒体、公共知识分子、公共道德与秩序的立场看无疑会引发各种的批评和指责，更多的反思肯定是那些持续了近一个世纪的有关国民性与国民素质的讨论，这些居民也必然是被指责为没有公共道德的"刁民"。这种他者视域下的中国现代性话语在今天仍是主流意识形态的重要组成部分，并由此构成了现行秩序维护者的集体无意识。诚然，作为项目设计者和项目实施的参与者，最终呈现效果与初始设计的落差并非我们所愿，但指责与批判并不能改变这一现实，而重新反思设计思路本身也许更具有现实性和在未来改进上的可操作性。

1. 设计者的想象与现实的冲突

首先，从委托方的立场看，如果想实现其规划的意图，也即保留这一区域内的"历史性建筑"，恢复在过去半个多世纪内被居民们"私搭乱建"所破坏的老建筑和街道的原貌，那么，首要任务就是"去贫民化"与引进"中产阶级"的生活方式以实现对低收入人口的"换血"目标。但由于这一规划"只迁不拆"的客观限制，使得生活于这条街道内的居民在搬迁事宜上有了更多的溢价空间。出于各种考虑，多数居民仍选择了继续生活在这里。部分腾退出的房屋被改建成了各种受到"中产阶级"、文艺青年以及外来游客青睐的咖啡、餐厅、画廊、书店、时尚小店等。于是就有了杨梅竹斜街现在的这种景象，即不同的人群、不同的生活方式、不同的符号、不同的时间性和空间性被压缩、并置、混搭在一条长约 500 米的街道内。

其次，站在设计人员的立场看，他们原初参与到这一项目的动力来自某种"文化使命感"以及这一代知识人共有的"怀旧情结"，他们希望通过设计去实现某种心目中的理想景象，即那种基于现代化条件下的，既清洁有序，又保持传统风貌的，既保留本地居民"原生态"的生活习俗，又有具备现代文化品位的休闲生活场景。但是，现实中居民生活的"原生态"在面对这种新环境的时候，他们的反应并不是欣然接受或融入其中，而是我行我素，保持着原有的生活状态与行为习惯。

最后，作为这一街道中的"主人"，从那些在此居住生活了几十年甚至更久的居民们的立场来看，生活环境的改变并没有给他们带来多少实惠。与他们自己习惯的生活方式、文化趣味、作息时间相比，那些新迁入的邻居们（咖啡馆、画廊、书店等）完全生活在另一个世界。尽管他们共存于同一条街道，互为邻里，但无论哪方面，他们之间都没有任何的交集。每当文化机构或商业公司出于推广目的而举办各类文化、商业活动的时候，本地居民的生活也只是作为这类活动的背景而存在。因为，任何一项活动都是围绕着那些新邻居们展开的，与他们的现实生活毫无关系。他们虽然是这条街道的主人，但却被那些"外来者"视为他者。作为新邻居们，那些咖啡馆、时尚商店、书店等展现给他们的客户（来访者、消费者、游客）的是：在一个破落的但有文化历史痕迹可循的背景中的"新视觉"，以及这些"新视觉"的载体——商品。就如居伊·德波在《景观社会》一书中开篇所言："生活本身展现为景观的庞大堆聚。"[1] 其特征是符号价值超过使用价值，展示价值大于交换价值。与之相比，那些作为背景存在的本地居民所

---

[1] Guy Debord, *The Society of the Spectacle*, New York: Zone Books, 1995.

展现出来的倒是生活本身。在这种新的视觉冲击面前，这些以中老年人口为主的社会底层居民与被称作"新新人类"的年轻一代相比倒是显得异常的淡定。这种淡定并非来自某种文化抵抗，而是因为，这是一个在中国现代化大潮中被边缘化的群体，是主流社会期望以最低的成本使他们尽早从视线中消失的群体，是京城坊间流传的贫富贵贱等级中最末端的群体。在这样的境遇下，这一群体所能有的反应或抵抗就只有他们的态度和行为了，那种我行我素的态度和那种被公共舆论认为的"刁蛮"的行为。

2. 观察与思考

委托方的意图、设计方的理想与本地居民们的我行我素共同构成了杨梅竹斜街现在这样一个现实图景，这图景就像福柯在其演讲录《另类空间》中对异托邦第三个特征所描述的："异托邦有权力将几个相互间不能并存的空间和场地并置为一个真实的地方。正是这样，在长方形的舞台上，剧场将一系列互不相干的地点连接起来。"[1]在杨梅竹斜街这个舞台上，如果观众们用一天的时间去观察这里所发生的事情，就会发现一幕幕有趣和耐人寻味的场景：

---

[1] [法]福柯：《另类空间》，王喆译，《世界哲学》2006年第6期。

场景一：
Scene one

早晨 8 点，在街边一个带有明显美国风格的小木屋门前摆放着北京人习以为常的早餐桌，不时有人坐在此处吃着豆浆、油条。往来于就餐者眼前的是手提塑料袋、穿着睡衣拖鞋的主妇，或拉着盛满蔬菜小车的老人，他们是刚刚从附近早市买菜回家的街道居民；下午 4 点以后，这间小木屋开始营业。这是一家专门出售美国旧式军服的小店，但却少有客人光顾，而贴着海报的玻璃上反射的是对面两家正准备收摊的由本地居民经营的廉价古董店。那座"美式"小木屋及其出售的商品是现实生活中只被少数年轻人青睐的对象，店主人为什么会选择此地作为其卖场不得而知。但它在此存在的象征性无疑大于其商业价值，因为在这条街道中确实不存在其目标客户，它的存在似乎更多地体现着一种姿态，即那种以所谓时尚、先锋、异类等为标志的姿态。但是，从那些市井小民对此的反应看，这一标志性的存在物似乎就是一个"异物"，其语义学上的意指功能也不在他们的符号认知系统内。就在这一不大的空间内，在早晨、夜晚不同的时间段落里，他们各自展现着自己不同的生活，各自都以碎片式的存在成为彼此的背景。

第一部分　方案实施

明显带有美国风格的小木屋前，摆放着最常见的小餐桌，不时有人坐着吃早饭。
In front of an American-styled cabin, a common-looking small table has been set up and from time to time people sit there eating breakfast.

011

场景二：
Scene two

下午两点以后，在一间街道居民娱乐用的公共棋牌室内，昏暗的灯光下，十几个中年男女围坐在几桌麻将桌前打牌聊天，街道上过往的行人、游客对他们似乎没有任何影响，但他们不喜欢被外来者拍照。与棋牌室相邻的是一栋民国时期的老建筑，建筑物前的信息牌提醒人们，这里曾是由某人创建于1917年的"世界书局"旧址，现为民宅。

居民的公共棋牌室与书局旧址标牌，一个是居民自发娱乐的场所，一个是知识圣地的象征；一个是当下的时间，一个是历史时间的片段；一个是与其招牌相符的实体空间，一个是留给人们想象的虚幻空间。无论是环境杂乱的棋牌室，或是环境典雅的"模范书局"，出入其中的人所进行的休闲活动的品质和性质似乎是匹配的，而出入于这两个不同空间的人们也各属于他们在社会空间中的不同位置。棋牌室内的居民们对游客拍照的排斥或拒绝，表明他们懵懂之中似乎明白这种拍照无论是作为"民俗"还是"低俗"的记录，他们的形象都会被表征为主流社会的"他者"。

第一部分　方案实施

紧邻 1917 年建造的"世界书局"旧址的灯光昏暗的棋牌室里，十几个本地居民在打牌聊天，来往的游客对他们似乎没有任何影响，但他们不喜欢被外来者拍照。
In the dimly lit chess and card room next to the site of the World Book Publishing House, which was built in 1917, dozens of local residents play cards and chat together. Visitors seem to have no influence on them, but they do not appreciate being photographed by outsiders.

场景三：
Scene three

上午十点，一家装修精致的家用陈设品商店的门上悬挂着英文"Closed"的标牌，橱窗里展示着各式设计精美的家用器皿，橱窗的落地大玻璃窗上反射的是对面居民正在"扩建"施工中的二层自住房。正在施工中的居民似乎并不顾及政府和商业公司为这条街道所设想的"美好蓝图"和改善居住环境的"良好愿望"，更不会想到与对面的时尚商店以及邻近的"新邻居"们在外观和形式上的协调，而是在他们力所能及的条件下扩展自家的生活空间。他们淡定自若的神情表明他们并不在意过往游客们投来的奇异目光，而对面时尚商店所展现的一切都是为了吸引游人们的目光。前者仅仅是为了生活本身而生产，而后者生产的是一类生活的形式。

场景四：
Scene four

某个由商业公司举办的公共活动日的下午，离杨梅竹斜街入口不远处，在一辆改装设计过三轮平板车前，一个姑娘正在向过往的游客推销自制的芝士蛋糕和小甜点。在她的对面是一家本地居民平日里光顾的"主食厨房"，门前摆放着北京人日常食用的大饼、烧饼。前方30米外，在一

第一部分　方案实施

一家装修精致的家用陈设品商店的橱窗上反射的是对面居民"扩建施工"中的二层自住房
In the window of a well-decorated household furnishings store, there is the reflection of a two-storey residential house opposite the shop that is in the process of undergoing "extensions"

015

家时尚小店的门前，一组为公共活动特殊设计的街边小凳上，一个住在对面大杂院里的大妈与一对前来旅游的情侣相邻而坐，不时有衣着光鲜时尚的年轻男女游客和上身赤膊的本地居民从他们面前走过，而大杂院门口坐着的老头儿和他身边的自用三轮车似乎表明了他的日常生活与眼前发生的一切毫无关联。出售芝士蛋糕的姑娘和其他的"时尚小贩"们都是由这一公共活动的主办方特邀而来的。从他们所销售的小商品看，目标顾客肯定不是本地居民，而是通过主办方的推广宣传慕名而来的各路到访者及游客。与每年在此举办的"北京设计周"活动一样，公共活动的主体与观众都是"外来者"，而这条有着"曾经的文化历史"的街道与生活在这条街道里的居民仅仅是作为公共活动的舞台布景。这一结果与委托方和设计方的初始意图不无契合，即搭建一个公共艺术活动的平台。但是，身处这一平台的现场也许会有另外一种感受，那就是无论这一活动中的卖方或是买方，艺术家、设计师或是慕名而来的观众更像是在"相互表演"，而被作为活动背景的、自由自在、无拘无束的街道居民则更像是看客。尽管这类公共活动对他们毫无现实的益处，那些对"外来者"们有效的视觉冲击似乎对他们也不发生作用，更谈不上所谓的审美提升。但是，在这种互为他者的境遇中，他们并没有丧失在这一公共场合里作为自身群体的主体性。

第一部分　方案实施

某个举办公共活动日的下午，本地居民与游客的生活在同一空间交织
On the afternoon of a public activity day, the lives of the local residents and tourists become interwoven in the same space

场景五:
Scene five

在街道中一个相对整洁的住户门前的花池旁,两名女游客正在一丛观赏花前相互用手机拍照。这类观赏花卉是园林单位为了美化环境栽种的,但这些专门为每户居民门前设计的花池中的花草、竹子、灌木等在整条街道中已经所剩不多,除了那些"新邻居",多数的花池中都被居民们依照他们自己的喜好栽种上了各种自养的花卉,甚至是食用类的丝瓜、豆角、小葱等植物。图片中的两位女士以观赏花为背景相互拍照,表明此处的花并不属于她们,而只属于她们照片中的背景及其某种象征意义。同样,这些观赏花也不属于当地的居民,因为,这些为了美化街道而栽种的花卉,主要是为了使外来游客能有一个赏心悦目的环境,无论是植物的种类或是栽种的方式如何具有形式美感,都与多数本地居民的审美意象不相匹配。居民们对观赏植物的破坏以及改种自家带有实用目的的植物这一行为,不仅是对公共资源的破坏,也是对所谓非功利性审美原则的挑战。然而,在居民们的深层意识里,所谓的公共性是不包括他们在内的,或许他们有所意识到某种外来的公共性正试图驱除和剥夺他们的共同性。从居民们对自己所栽种植物的细心程度看,他们似乎并不是不喜欢花草,而是因为他们有属于他们这一群体自己的审美文化与传统,尽管这种审美活动并不属于非功利性的,纯粹的审美范畴,但这

第一部分　方案实施

两名女游客在观赏花池前互相拍照，而本地居民更喜欢在花池中栽种自养的花卉或豆角、小葱等蔬菜。
Two female tourists take pictures of each other in front of ornamental flower beds, however the local residents prefer growing their own flowers or even vegetables such as beans and scallions in the flower beds.

同样是一种感性经验。这种感性经验在现代性进程开启之前早已渗透在中国百姓的日常生活之中。

如果以美观与丑陋、传统与现代、文明与野蛮、高雅与粗俗、知识与愚昧等二元对立的惯性思维以及基于这种思维下的趣味好恶去感受、理解和阐释发生在这个舞台上的一切，那么，这一充满异质性的舞台空间在线性时间观的延伸中向着同质化的方向发展，最终成为仅仅被消费的"文化标本"或曰"拟像"就是其必然的结果，也即那种被重新编码的所谓"老北京文化"。当柴米油盐、衣食住行不再是真实的生活的时候，那么，怀旧的想象与未来的幻象就将填充它的全部意义，被遮蔽的恰恰是在"过去"与"未来"之间的当下。而这一结果并不是设计者的初衷。

3. 设计与设计之外的社会实践

与多数发展中国家一样，经济发展与城市化进程带来的另一社会现象就是：城市中存在着数量庞大的社会边缘人群和边缘区域。在北京，被称为"大杂院"的居住区内，生活着大量的低收入市民以及外来流动人口。外来务工人员的涌入与城市人口的自然增长超过旧城区的城市功能承载能力，导致旧城区的居住环境持续恶化，对生存空间的扩张更是改变了北京古城原有的院落建筑格局。这些看似贫民区的"大杂院"就散落在有着 600 年历史的北京老城区内,这些居民虽然居住在城市的中心,但他们的处境却属于当下社会的边缘。破败与杂乱的居住景象覆盖在北京旧城的城市肌理之上,与坐落于城市外围的、新兴的 CBD 区域和"高尚住宅区"形成了鲜明对比。

空间关系的变化所折射出的不仅是居民物质生活差别的扩大,更是社会关系的改变。原本基于家族、家庭与"单位"的"熟人社会"逐渐成为了历史,取而代之的是孤立的个人与分散的独立家庭的碎片式存在。持续的经济开发和快速的"城市改造"使旧城区的人们生活在一种不稳定的状态中。在拔地而起的商业设施与新城区的映照下,旧城区居民无力改善自己的生存环境,他们所在的熟悉的街区,成了被"发展"遗忘的角落。窘迫的生存境遇重新塑造了他们与所在城市的关系,瓦解着他们对于所在街区的认同感。部分原住居民的迁出与外来人口的迁入,使曾经相互熟悉的街坊邻里关系逐渐消失。陌生感与缺乏信任使得这里的人们相互疏离,彼此漠视。

在过去的 40 年里,快速的城市化进程不仅改变了一座城市的外观,同时也改变了生活于城市中的人们之间的社会关系。经济现代化所特有的劳动分工体系、分配体系在创造了经济奇迹的同时,也不可避免地带来原有社会结构的解体和与之相关的社会危机等一系列现代性的后果。这一后果在当下中国的城市和乡村所显现的情况之一就是：原有的熟人社会结构正在逐渐解体，"陌生人社会"正在形成。这一社会原子化的趋势使得原有熟人之间的默契与协商中介逐渐消失，而市民社会的"公共性"秩序尚未形成，这必然导致社会转型时期人际关系的疏离与相互信任的缺失。个体自由的获得也伴随着个人主义的膨胀，导致道德失范。在这一社会现代性的转型过程中，以多种方式重建社会中间组织以形成新的人际关系的连接纽带，是解决由社会分化所带来的危机与挑战的应对之策，由此，才能缓解因社会的去组织化所造成的游离个体的认同危机、焦虑与孤独感；化解由于个体间的、邻里间的、社群间的互不信任而产生的矛盾与冲突。

以上这些社会问题在杨梅竹斜街都有其现实的表现，不论是改造前或改造后。街道外观的改变并不能解决深层的结构性问题，但结构性问题却会消解甚至扭曲对外观改变的种种努力。针对这些问题，设计团队在整体工程完成之后进行了一系列社会调研，通过挨户走访以了解居民们的真实想法和愿望。在调查中我们发现了一种普遍的现象，即居住在这里的人们大多喜欢在自家狭窄的空间缝隙中，以随手即拾的容器种植自己喜欢的花草和日常食用的蔬菜。这种自家种植的花草与蔬菜在外观上也许不如公共绿化的植物美观，种植所使用的器皿更是五花八门，废弃的泡沫盒、脸盆等日常生活的废弃物都被作为容器。这一现象说明，有限的物质条件可以使他们忽略对外观的追求从而因陋就简，但人们对日常生活的趣味并没有因此而减少。对这些生活于社会底层的人们，花开花落，植物的生长与收获也许就是一种心灵的慰藉，一种对现实的默认中的精神寄托。

根据前期的调研结果，设计团队在 2015 年北京设计周期间开启了一项名为"胡同花草堂"的公益项目。该项目针对的是生活于杨梅竹斜街 66—76 号院夹道里的五户人家。夹道是一条长 66 米，最窄处仅 1 米、最宽处不足 4 米的通道，是百年院落与持续扩张的居住空间在数十年间挤压中形成的通道，由五户居民共用，其中包括定居在这里 400 余年的老北京家庭、搬来 20 年左右的新原住民，以及暂住的外来打工者。

此项目旨在凝聚居民们共同的生活情趣与精神追求，以建立共享"花草堂"的方式介入这一夹道的空间改造中，为五户彼此相邻却相互疏离的居民建立邻里间有效的交往模式，缓解在急速城市化进程中由于社会中间组织的缺位所造成的人际关系的离散，从而使暂居于此的人们能够通过养花、种菜这种自发的中介形式相互交流，互通有无，并形成某种新的、约定俗成的公共秩序。共同的爱好与向往所形成的这种空间形式的中介，不仅可以使游离于社会底层的个人、家庭之间建立某种认同与沟通，也可缓解由于"暂住"而产生的焦虑感，并通过共建与共享"花草堂"这一连接中介，找到自己的归属感从而"安住"于此，相续共生。从实验项目的规模上看，这仅仅是一条狭长曲折的通道，所涉及的人口有 20 人左右，而能够用于"花草堂"建设的空间不足 10 平方米。与城市广场、公园、艺术区等相比，这种最小尺度的公共空间设计也许不具有推广价值，但其不可化约的特定性正是这一实验的意义所在。我们并没有预设我们的实验必然成功，仍有许多不确定因素是我们难以把握的，其结果还需要长时期的观察与不断的调整才能显现。

"胡同花草堂"的项目入选 2016 年威尼斯建筑双年展中国馆的展览。作为展览的一部分，展出期间，夹道中的五户居民通过视频直播与远在千里之外的各国观众进行了互动。2016 年和 2017 年北京设计周活动期间，设计团队又相继推出了"日常生活 VS 景观设计""种瓜得瓜，种豆得豆""三岁，胡同花草堂"等主题活动。

相比于团队曾经从事的景观设计项目来讲，这些公益项目更接近一次社会学意义上的社会工作实验。我们努力的目标不是设计一个供外来者或旅游者观看与消费的景观，而是通过设计参与到当代中国的社会重建之中。虽然，经济利益驱动与政治诉求是一个地区发展与文化变迁的基础动力，但是这些来自外部的力量如不能与被改造的对象有机地结合与良性的互动，其结果必然是破坏性的。我们所关注的是在社会转型期如何提升个人的尊严，重建个体间、群体间相互尊重、和谐共生的人际关系。我们所努力的方向就是在现有的条件提升本地居民的自主性和参与性，通过各种活动使他们那些如种植、捡拾废品等自主行为、个人行为变成受尊重的公共行为。由此而引发的思考是：如何定位设计与设计者在社会生活中的角色与位置。我们参与公益项目并不是出于展示自身的社会责任感或同情之心，而是通过各种尝试去发现并建立一种有效的机制，从而使那些生存于社会边缘的人们从中受惠。在这一过程中，设计者必须在自己的审美理想、设计理念与普通民众的生存逻辑、审美趣味之间做出妥协，让设计回归到它应有的功能，发挥它最原初的潜质。在设计实践中，妥协就意味着设计者需要放弃自己的某些主观想象与意志，甚至是理想，而仅仅作为一个社会生活与经济发展中的协调者。这一角色的转换似乎消解了设计者的主体性，也弱化了"设计师"曾有的光环。但是，设计，不论是建筑设计或环境设计都应该是针对具体设计对象的，因时、因地、因人而异的解决方案。对特定情境中的人的关照应该是设计的核心，也是目的，而形式与风格仅仅是这一目的的外化。

本书内容以杨梅竹斜街改造项目为核心，旨在展现设计团队在过去的 6 年时间中为这一长约 500 米的老城街道的改造与提升所付出的努力，以及这些努力的背后对老城区改造所进行的思考。

童岩，中国人民大学艺术学院

2018 年 4 月 于北京

Summary

The organic renewal project of Yangmeizhu Xiejie is part of the renovation project of Beijing's old city. It is a project led by the local government and operated on the basis of the commercial model provided by the developer. The expected goal of the project can be defined as: on the basis of avoiding the destruction of the texture of the old city streets, to transform the infrastructure and the aesthetic appearance of the streets, introduce new business mechanisms through housing replacement, and building this old street into a sustainable and self-generating tourism product whilst at the same time giving priority to local residents. From project planning to project implementation, the government departments, developers, architects, landscape architects, street residents who involved in the project are involved based on their own respective goals, benefits and interest. Judging from the effect of the completed project, it is undoubtedly the result of compromise amongst the disparate parties.

安住·杨梅竹斜街改造纪实与背后的思考

Compared with the nearby Dashilan area, the Yangmeizhu Xiejie has become much quieter after the renovations. There are no local shops, restaurants or souvenir shops, and few visitors featuring different accents or skin color. Apart from a short period during which there is noise caused by public cultural events such as the Beijing Design Week, which occurs a few times a year, everyday life here is quiet and peaceful. Residents' gossip is mostly spoken in the old South Beijing accent.

With regards to the overall appearance of the street based on the completed transformations, aside from the specially designed pavement, the steps and well caps feature the word "Dashilan" across the street, which is about 500 meters long, and the several distinctive fashion shops, cafes, bars and bookstores, it seems that little has changed as relating to the general character of the street, and there are still parts that are slightly cluttered. From the point of view of greenery and beautification, besides the potted plants and flowers placed by the municipal gardening department during the Beijing Design Week and on major holidays, during the normal weekdays Yangmeizhu Xiejie even appears to be a bit "desolate". The plants in the house-side flower pools have been carefully designed by the designers for greening and "beautifying" the street environment, however now these scenes have changed beyond recognition. Some of the flower pools have even been replaced with the residents' own towel gourds and shallots. From the design point of view, the reconstructed streets are quite different from some of the famous "old streets" or "art districts" in major cities in China, in the sense that there are not even a few spots that would be suitable for taking a picture. However, this 400 meters area of pavement and local housing renovation took nearly three years from the planning stage, to design stage to the completed construction. A lot of work was expended in order to coordinate the overall style of the street design in accordance with the benefits and interests of the street residents.

As mentioned earlier, the transformation of the Yangmeizhu Xiejie is the result of the conflict and compromise between the intentions, objectives and benefits of all the parties involved. Among them, the most noteworthy and deserving of our reflection is the attitude of the local residents towards this public project. In accordance with the requirements of the client, the designer has adopted pressure-resistant floor tiles constructed by the means of traditional crafts and created modules in the design of street floor pavement. The purpose is to create a connection with some of the old buildings that are still in their original state, and to more effectively bear the pressure of automobiles. Because the original state of greenery was very poor, with almost no trees to speak of, the overall intent of the greenery plans was to increase the greening rate of the street by adding significant amounts of flower beds that could feature shrubbery and may be connected with the residential houses. However, all of these initiatives to beautify the environment were not supported by the majority of the residents. According to the construction record: after construction, the plants that were planted in the flower beds needed to be replaced frequently, because these plants were often excavated or destroyed by the residents, and the residents even planted their own plants in these flower beds for their own practical purposes, such as shallots, towel gourds, beans, etc. In addition, according to municipal greening units, local residents take 20% of potted plants placed during various public events back to their individual homes.

The above negative information will undoubtedly arouse criticism from the viewpoints of the public media, public intellectuals, and with regards to public morality and order. Upon further reflection, it is unavoidable to unearth discussions that have churned on for almost a century regarding the national character and values. It is no surprise that these inhabitants have also been accused of being "tricksters", devoid of public morality. Indeed, the discourse of Chinese modernity from the perspective of the other is still an important part of the modern-day mainstream ideology, and thus constitutes the collective unconsciousness of the maintainers of the current order. Admittedly, as project designers and participants, the difference between the final presentation and the initial design is not in line with our intentions. However, to criticize and repudiate these results has no bearing on whether or not we can change this reality. Rather, we must approach the design itself by reconsidering if the design may be more realistic and feature greater operability as relating to future improvements.

The Conflict Between the Designer's Imagination and Reality

Firstly, from the viewpoint of the client, if we want to achieve the original purpose of the planning, which is in essence to preserve the "historic buildings" in the region and restore the old buildings and streets that have been destroyed by the "private construction" carried out by residents over the past half century, then the first task is to "dispel of the poverty" and introduce a "middle class" lifestyle in order to achieve the goal of applying a "makeover" for the low-income population. However, due to the objective restrictions of the "relocation without demolition" policy, the residents living in this street experienced a premium on the surrounding properties. For a variety of reasons, most of the residents still chose to continue living here. The partially vacated houses were converted into cafes, restaurants, galleries, bookstores, boutique shops, etc., which are favored by the "middle class" and artistic young people as well as foreign tourists. Thus, the scene in Yangmeizhu Xiejie nowadays, is one of disparate groups of people, carrying out different lifestyles, with different symbols, and different concepts of time and space, which are compressed, juxtaposed, and combined in a street that is approximately 500 meters long.

Secondly, when considering the position of the designer, their initial involvement in the project was motivated by a sense of a "cultural mission" and the "nostalgia complex" of this generation of intellectuals, and their hope was to achieve some sort of ideal vision through design. Essentially, the designers concept was based on the idea of clean and orderly modern conditions, whilst also maintaining the traditional style, and retaining the "original ecology" of the local residents lifestyle and customs, and at the same time providing scenes of leisure life with a modern cultural taste. However, in reality, when the "original ecology" of the residents' lives came into contact with this new environment, their reaction was not to accept these changes or blend into this new environment, but to follow their own course and maintain their original living conditions and behavioral habits.

Lastly, as the "owner" of this street, from the perspective of those who had lived here for decades or even longer, the change of living environment did not bring many particular benefits to them. Compared with their customary lifestyle, cultural interests and leisure time, those new neighbors that had moved in recently (ie. cafes, galleries, bookstores, etc.) are living in a completely separate world. Though they are neighbors that live together in the same street, regardless of this, there is no intersection whatsoever between them. Whenever a cultural institution or a commercial company hosts cultural events or business activities for promotional purposes, the lives of the local residents are only present in the context of such activities. This is because all of the activities revolve around the new neighbors and have nothing to do with the day-to-day lives of the original residents. Although they are the owners of this street, they are regarded as others by the "outsiders". As new neighbors, those cafes, fashion stores, bookstores etc. want to present themselves to their customers (visitors, consumers, tourists) in a certain way: a "new vision" set against the backdrop of a run down alley, yet one in which it is possible to trace the historical and cultural roots. In addition, this "new vision" is conveyed via a particular means - commodities. As mentioned in Guy Debord's opening in the book The Society of the Spectacle, "…all of life presents itself as an immense accumulation of spectacles…" Its characteristic is that the value of the symbol exceeds the value of its use, and that the display value is greater than the actual exchange value. In contrast, it is the original residents functioning in the background that actually embody the genuine lifestyle of the alley. In the midst of this modern visual impact, these middle-aged and old-aged people who live at the bottom rungs of society are unusually calm when compared with the younger generation, that are seen as the "new people". This type of calm assuredness does not come from a certain cultural resistance (namely, what the European "situational" people did during the era of Guy Debord), rather, it is because they are a marginalized group caught amidst the tide of China's modernization; they find themselves as the lowest class in the society, and the mainstream society expects them to disappear from sight as soon as possible at the lowest cost possible. In such a situation, the only response or resistance that this group can muster is found in their attitude and behavior; conjuring a way of acting in an "unruly" manner as seen by public opinion.

Observation and Thinking

The intentions of the client, the designer's ideals and the local residents' actions constitute the picture of present-day Yangmeizhu Xiejie. This scene is strikingly similar to the description of the characteristics of Heterotopia in Focault's speech "Dits et écrits": "Heterotopia has the power to juxtapose several mutually exclusive spaces and venues into a real place. In this way, on a rectangular stage, the theater connects a series of irrelevant places together." With regards to the stage of Yangmeizhu Xiejie, if an audience were to spend a day watching the occurrences in the locality, they would find themselves witnessing a variety of interesting scenes.

Scene 1: 8 o'clock in the morning on the street. Traditional Beijing style breakfast tables are placed in front of a small wooden cabin that has been clearly constructed in an American style. Now and then people sit here drinking soybean milk and eating fried bread sticks. In front of the diners are housewives wearing their pajamas and slippers, and carrying plastic bags, as well as elderly men pulling carts full of vegetables. They are residents of the street who have just bought vegetables from the early-morning markets nearby. After 4 p.m., this wooden cabin opens.

It is a small shop specializing in the sale of old style military uniforms from the United States, but few customers patronize it. In contrast, in the reflection of the poster-covered windows two cheap antique shops across the street run by local residents are finishing up for the day. In reality, the "American" cabin and the goods it sells only generate interest from a small number of young people, and we don't really know why the shop owner chose this place as his store. Nonetheless, its symbolic presence is undoubtedly greater than its commercial value. As there's no target customer which exists in this street, the very existence of the store seems more akin to that of a gesture, namely, a gesture marked by the so-called fashion, pioneering spirit and heterogeny. However, judging from the reaction of the people in this area, this iconic existence appears to them a "foreign object", of which the semantic function of this object is not present in their symbolic cognitive system. In this limited space, operating in different time periods whether be it the early morning or late evening, different people are carrying out vastly different lives, and each group has become a background of the other within a fragmented existence.

Scene 2: After 2 p.m., in a public chess room which serves as entertainment for the local street residents. Under dim lights, more than a dozen middle-aged men and women sit around a few mahjong tables playing cards and chatting. Passing pedestrians and tourists in the street appear to have no influence on them, but they don't appreciate being photographed by outsiders. Adjacent to the chess room is an old building from the Republic of China era. The sign in front of the building reminds people that it was the site of the "World Book Publishing house", founded by an individual in 1917, and is now a residential building. The residents' public chess room and sign commemorating the publishing house: one is for the residents to enjoy spontaneous entertainment, the other is a symbol of a sanctuary of knowledge; one exists in the present, whilst the other is a fragment of history from another period. One is of a physical space that is consistent with its signboard, whereas the other is an illusory space for people to exercise their imagination. Whether it's a cluttered chess room or an elegant "model bookstore", there is a consistency in the quality and character of the leisure activities carried out by the people who go in and out of these facilities, which clearly displays the different respective positions of the people in this social space.

Residents in the chess room dislike or even reject being photographed by tourists, suggesting that they seem to understand that such photographs, regardless of whether they are recorded as "folk" or "vulgar", will be characterized as the "other" by the mainstream society.

Scene 3: At 10:00 a.m., an English "closed" sign hangs on the door of a well-decorated home furnishings store. Exquisitely designed household utensils are displayed in the full-length windows, and the large glass windows reflect the two-story houses directly opposite that is in the process of being "extended" by the residents. The residents that are in the process of these renovations seem to have no regard for the "magnificent blueprint" and "good intentions" of the government and commercial companies with regards to the street, not to mention any sort of coordination or harmony with the fashion stores across the street that are the "new neighbors". Rather, they are expanding their own living space under the conditions provided for by their own ability. Their calm expression shows that they do not care about the strange sight of tourists passing by. By contrary, everything the fashion shop across road does is in order to attract the attention of passing tourists. One simply produces for the sake of life itself, while the other is attempting to produce a particular style of life.

Scene 4: An afternoon during which a business company has organized a public activity day. Not far from the entrance of Yangmeizhu Xiejie, in front of a converted tricycle, a girl is selling homemade cheesecake and desserts to passers-by. Across from her is a "staple food kitchen" frequented by local residents, which features stacks of pancakes and buns that have been placed in front of the door. Thirty meters ahead, in front of the doorway of a fashion shop, there are a group of street-side stools that have been specially designed for public events. An auntie living in the compound opposite sits next to a tourist couple, whilst fashionable young male and female tourists and half naked local residents walked past them every now and then. An old man sitting at the entrance to the compound with his tricycle beside him seems as if his daily life has nothing to do with any of the events that are happening. The girl selling cheesecake and the other "fashion vendors" have been invited by the organizers of the public event. Based on the merchandise they sell, their target customers are certainly not local residents, but rather the visitors and tourists from different places that have been attracted through the promotion of the sponsors. Like the Beijing Design Week, which is held here every year, the main body and audience of public events are "outsiders". This street with its "old cultural history" and the people living in it merely function as a stage set up for public events. This result is consistent with the original intention of the client and the designer: to build a platform for public art activities. However, there may be another feeling that exists on this stage; whether it is the seller or the buyers involved in the activity, artists, designers, or spectators, they are more likely to be "acting together". Instead, the street residents in the background are simply being themselves, and in this way they are more like the audience. Although such public activities are of no practical benefit to them, the effective visual impact that is felt by the "outsiders" seems to have no effect on them, let alone the so-called aesthetic promotion. However, in this situation of mutual otherness, they did not lose their subjectivity as their own individual group functioning in this public place.

Scene 5: Two female tourists use their mobile phones to take pictures of each other in front of a cluster of ornamental flowers near a flower bed in front of a relatively tidy residential door located in the street. These ornamental flowers have been planted by landscaping units in order to beautify the environment, but there are few flowers, bamboo, shrubs, etc. that are still remaining along the entire street with regards to this greenery which had been specially designed for each household. Apart from those nearest to the "new neighbors," most of the flower beds have been replanted by the residents in accordance with their own tastes, the greenery being replaced by flowers the residents have grown themselves, and even gourd, beans, shallots and other vegetation. The two women in the picture photograph each other against the backdrop of ornamental flowers, indicating that the flowers here do not belong to them, but rather to the background itself, which possesses some symbolic significance in their photographs. Likewise, these ornamental flowers do not belong to local residents either, because they have been planted for the purposes of street beautification, and mainly in order to provide a pleasant environment for tourists that come from outside. No matter with regards to the type of plants or methods by how they are planted, they are not in accordance with the aesthetic image of most local residents. The destruction of ornamental plants by the inhabitants and the transformation of replanting their own plants for practical purposes is in essence and action that not only destroys public resources, but also challenges the so-called non-utilitarian aesthetic principles. Nonetheless, existing in the deepest sense of those residents is the reality that these so-called public features are not inclusive of them. Perhaps they realize that some form of external publicity is trying to drive them away and deprive them of their commonality. Judging from the degree of carefulness that they apply towards their own plants, it does not seem that they dislike flowers and plants. Rather, it is because they have their own aesthetic culture and tradition, and although this aesthetic activity does not belong to the non-utilitarian or purely aesthetic category, it is also a kind of perceptual experience in itself. This perceptual experience has already penetrated into the daily life of the Chinese people long before the process of modernity started.

If we sense, understand and interpret everything that happens on this stage through the prism of binary opposition and conceptual thinking based on the taste of beauty and ugliness, tradition and modernity, civilization and barbarism, elegance and vulgarity, knowledge and ignorance, then this space which is full of heterogeneity, is developing towards the homogenization when witnessed through the extension of linear time. The inevitable result is that ultimately, it will become a "cultural specimen" or "simulacrum" that is for consumption. When the daily necessities of life are no longer at the core of our subjective reality, the nostalgic imagination and the illusion of the future will provide a fuller meaning. It is the present between the past and the future that is obscured. This result is not the original intention of the designer.

Design and Social Practices Beyond Design

Like most developing countries, another social phenomenon brought on by economic development and urbanization is: the emergence of large numbers of marginal social groups and fringe areas that exist in cities. In Beijing, the residential area known as the "mixed courtyard" is home to a large number of low-income citizens and migrants. The influx of migrant workers and the organic growth of the urban population has exceeded the capacity of the old city to bear the necessary urban functions, resulting in the continuous deterioration of the living environment of the old city. As such, the expansion of living space has changed the original courtyard architecture of the Beijing old city. These seemingly slum-dwelling "courtyards" are scattered around the 600-year-old urban areas of Beijing. Although these residents live in the center of the city, their situation is at the edge of the present society. Run-down and chaotic living scenes overlay the texture of Beijing's old city, contrasting sharply with the burgeoning CBD area and the "noble residential areas" situated on the outskirts of the city.

The change in spatial relationship reflects not only the enlargement of the difference present in the residents' material lives, but also the change existing within social relations. The "unit" based on family and "society of acquaintance" have gradually faded into history, replaced by fragments of isolated individuals and scattered independent families. Sustained economic development and rapid "urban renewal" have left people in the old urban areas living in an unstable state. Compared with the commercial facilities and new urban areas rising abruptly from the ground up, the residents in the old urban areas are unable to improve their living environment. Their familiar neighborhoods have become the forgotten corners of "development". Their embarrassing living situations have reshaped their relationship with the city, disintegrating their existing sense of identity within their neighborhoods. The relocation of some indigenous inhabitants and influx of immigrants from other provinces has resulted in the once familiar neighborhoods gradually disappearing. Unfamiliarity and a lack of trust have made people here alienated from each other and ignored by each other.

In the past 40 years, rapid urbanization has changed not only the appearance of the city, but also the social relationships amongst those people living in it. The labor division system and distribution system which are peculiar to economic modernization have not only created an economic miracle, but also inevitably brought about a series of consequences relating to modernity such as the disintegration of the original social structure and the subsequent social crisis. One of the consequences of this in both China's urban and rural areas is that the original social structure of acquaintances is gradually disintegrating, and a "stranger society" is forming.

This trend of social atomization has led to the gradual disappearance of tacit understanding and intermediaries for negotiation among the original acquaintances, while the "public" order of civil society has not yet formed. This inevitably has led to the alienation of interpersonal relationships and a lack of mutual trust amidst this social transformation; the acquisition of individual freedom has also been accompanied by the expansion of individualism, resulting in moral anomie. In the process of this transformation towards social modernity, the reconstruction of social intermediary organizations in a variety of ways in order to form new interpersonal ties is the solution to this crisis and the challenges that have been brought about by social differentiation. Only in this way can we alleviate the existing identity crisis, and the anxiety and loneliness of dissociated individuals which has been caused by social disorganization. In doing so, it may be possible to resolve the mutual distrust and conflicts among individuals, and between neighbors or entire communities.

All these social problems have expressed their realistic manifestations in Yangmeizhu Xiejie, whether before or after the renovations. The aesthetic changes applied to the street cannot solve the inherently deep structural problems. However, these structural problems may dispel or even distort the efforts of these aesthetic changes. In response to these problems, the design team carried out a series of social surveys after the completion of the project, by means of door-to-door visits in order to understand the real thoughts and wishes of the residents. In the survey, we found a common phenomenon; that people living here prefer to grow their favorite flowers and vegetables in discarded plastic containers which they have placed in their own personal narrow spaces. The homegrown plants and vegetables may not look as beautiful as the plants installed as part of the public greening, and the utensils used for growing them are more varied, such as discarded foam boxes, washbasins and other waste from daily life which are used as containers.

This phenomenon shows that under limited material conditions, this group can be influenced to neglect the pursuit of external appearance and remain simple. Despite this, their interest towards their daily life has not been reduced. For these people who exist at the bottom rungs of society, flowers bloom and wilt, and the growth and harvest of plants may be of a kind of spiritual comfort, a kind of spiritual sustenance that may be found in their default reality.

According to the results of the previous research, the design team launched a public welfare project called "Hutong Flora Cottage" during Beijing Design Week in 2015. The project targeted five families living along a narrow passageway of Yangmeizhu Xiejie in the No. 66 to 76 residences. The passageway is a length of 66 meters, of which the narrowest part is only 1 meter wide, and the widest part is less than 4 meters wide. This passageway has been formed by the continuous expansion of residential space in the previous decades. It is shared by five households, including old Beijing families who have lived here for more than 400 years and newer transplants who have moved into this space for approximately 20 years, as well as temporary migrant workers.

The purpose of this project is to unite the common interests and spiritual pursuit of the residents. To reform this space we created a shared "flower garden", establishing an effective mode of communication between the five residences who are alienated despite being in such close proximity to each other. To alleviate the dispersion of interpersonal relationships caused by the absence of social intermediary organizations due to the rapid urbanization process, temporary residents could be able to communicate with each other through the spontaneous intermediary form of flower cultivation and vegetable cultivation, and in the process form a new, customary public order. This spatial form of intermediary which has been formed by common hobbies and yearnings is not only beneficial with regards to helping individuals and families at the bottom rungs of the society in establishing some kind of identity and communication, but can also alleviate the anxiety caused by "temporary residence" and provide a space in which to find their own sense of belonging by co-building and sharing the "flower garden" as a medium for creating connections, so as to "settle down" and coexist in succession. Based on the scale of this experimental project, it is only a narrow and winding channel, involving approximately 20 people, and the space which can be used for the "flower garden" is less than 10 square meters. Compared with city squares, parks and art districts, this minimal scale of public space design may not be worth popularizing, but its irreducible nature is the significance of this experiment. We do not presuppose that our experiment will succeed, and there are still many uncertainties that we cannot grasp. The results will require long-term observation and constant adjustment.

The "Hutong Floral Cottage" project was selected as an exhibition for the China Pavilion of the Venice Biennale 2016. As part of the exhibition, five households in the alleyway interacted with audiences thousands of miles away via live video. During the Beijing Design Week events in 2016 and 2017, the design team launched themed activities such as "Daily Tao VS Landscape Design", "Together we sow, Together we harvest", "Three Years Old, the Hutong Floral Cottage", and so on.

These public welfare projects are more akin to a sociological experiment than the landscape design projects the team worked on before. Our goal is not to design a landscape for visitors or tourists to visit and consume, but to participate in the social reconstruction of contemporary China through design. Although economic interests and political appeals are the basic driving force for regional development and cultural change, if these external forces cannot be organically combined with the transformed object and result in a benign interaction, then the effect will be destructive. What we are primarily concerned with is how to enhance personal dignity in this period of social transformation, reconstructing interpersonal relationships with respect and harmonious co-existence among individuals and groups. Our direction is to enhance the autonomy and participation of local residents under the existing conditions through various activities that facilitate their autonomous and personal actions, such as planting and picking up waste, into respectable public actions. Thus the thinking is: how to position design and the designer's role and position in social life. Instead of showing our sense of social responsibility or compassion, we participate in public welfare projects by trying to find and establish an effective mechanism which is focused on benefiting those who live on the margins of society. With regards to this process, designers must make a compromise between their aesthetic ideals, design concepts and ordinary people's survival logic and aesthetic tastes, returning to the proper and intended function of design, allowing it perform its original potential. In the practice of design, compromise means that designers must be required to abandon certain aspects of their subjective imagination and will, including even their ideals, and function only as a coordinator of social life and economic development. The transformation of this role seems to dissolve the designer's subjectivity and weaken the aura that the "designer" once had. However, design, whether in architectural design or environmental design, should be targeted at specific design objects, and solutions that vary from time to time, from place to place, and from person to person. This should be the core and also the primary goal of design, in which the form and style are merely the externalization of this goal.

The content of this book is centered on the transformation project of Yangmeizhu Xiejie. It also aims to demonstrate the efforts of the design team with regards to the renovations and upgrades of this 500-meter-long old city street during the past six years, in addition to the careful thinking behind these efforts with regards to the transformation of old urban areas.

Tong Yan, School of Arts, Renmin University of China,
April 2018, Beijing.

二、杨梅竹斜街有机更新 (2012—2013)
Yangmeizhu Xiejie's Organic Renewal

杨梅竹斜街全长 496 米，为东北至西南走向，东北起煤市街，西端至北京城最短的胡同——一尺大街。在乾隆十五年（1750）绘制的京师全图中原名杨媒斜街，光绪或者民国年间，谐音加字雅化为杨梅竹斜街。

历史上，杨梅竹斜街曾为"龙脉交通车马辐辏之地"。金朝时，宣南一带曾是金中都城址所在，蒙元灭金后，在中都东北重建大都。由于元大都新城（北城）与金中都旧城（南城）之间的人货往来，年深日久便自发形成了若干条由西南斜向东北的捷径街道。杨梅竹斜街及其附近的铁树斜街（原李铁拐斜街）、樱桃斜街与棕树斜街（原王广福斜街）等都由此形成。

清末民初是杨梅竹斜街极为繁盛的时期，汇聚了北京四大商场之首青云阁，明清著名的翻译机构四译馆，以及众多书局商铺。

第一部分　方案实施

Yangmeizhu Xiejie is a 496-meter long street that runs from northeast to southwest. It starts from Meishi Street in the east and ends at Yichi Street, the shortest alley in all of Beijing, in the west. The Jing Shi Quan Tu (The Complete Map of the Inner Capital City Beijing), drawn in the 15th year of the Qianlong Emperor of the Qing Dynasty (1750), records that the street was originally named Yangmei Xiejie. However, during the Republican Era of China, it was changed to its current name, Yangmeizhu Xiejie, to evoke a greater sense of elegance. Historically, Yangmeizhu Xiejie was a crucial artery and hub for transportation; during the Jin Dynasty, Xuan Nan was the location of the imperial capital, however, after the Jin Dynasty was destroyed by the Mongols, they rebuilt the capital in the northeastern Zhongdu. Because of the frequent movement of people and goods between the new Yuan capital and former Jin capital, several diagonal shortcuts gradually appeared over time. Yangmeizhu Xiejie and its neighboring Tieshu Xiejie (formerly Litieguai Xiejie), Yingtao Xiejie, and Zongshu Xiejie (formerly Wangguangfu Xiejie) were formed as a result.

The late Qing Dynasty and early Republic of China were the most prosperous periods for Yangmeizhu Xiejie, which was home to the Qingyun Pavilion, the foremost of the four major markets in Beijing, the Translators' College (Siyi Guan), a famous translation organization from the Ming and Qing dynasties, and numerous bookstores and shops.

安住·杨梅竹斜街改造纪实与背后的思考

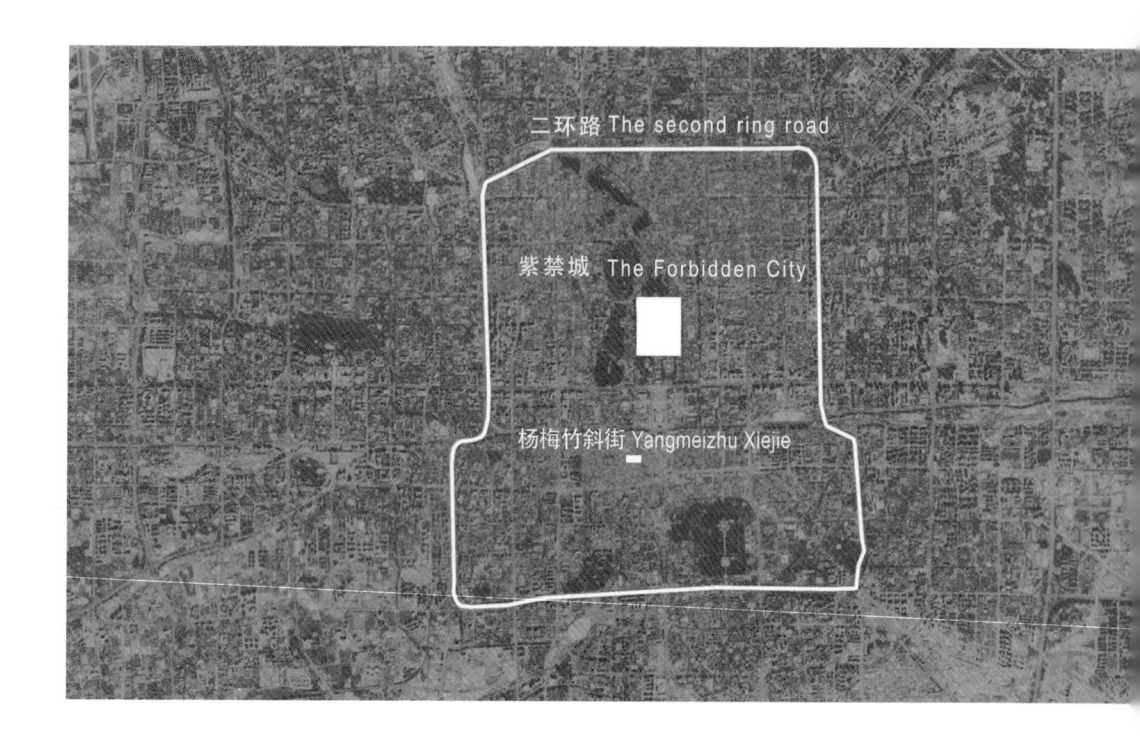

二环路 The second ring road

紫禁城 The Forbidden City

杨梅竹斜街 Yangmeizhu Xiejie

第一部分　方案实施

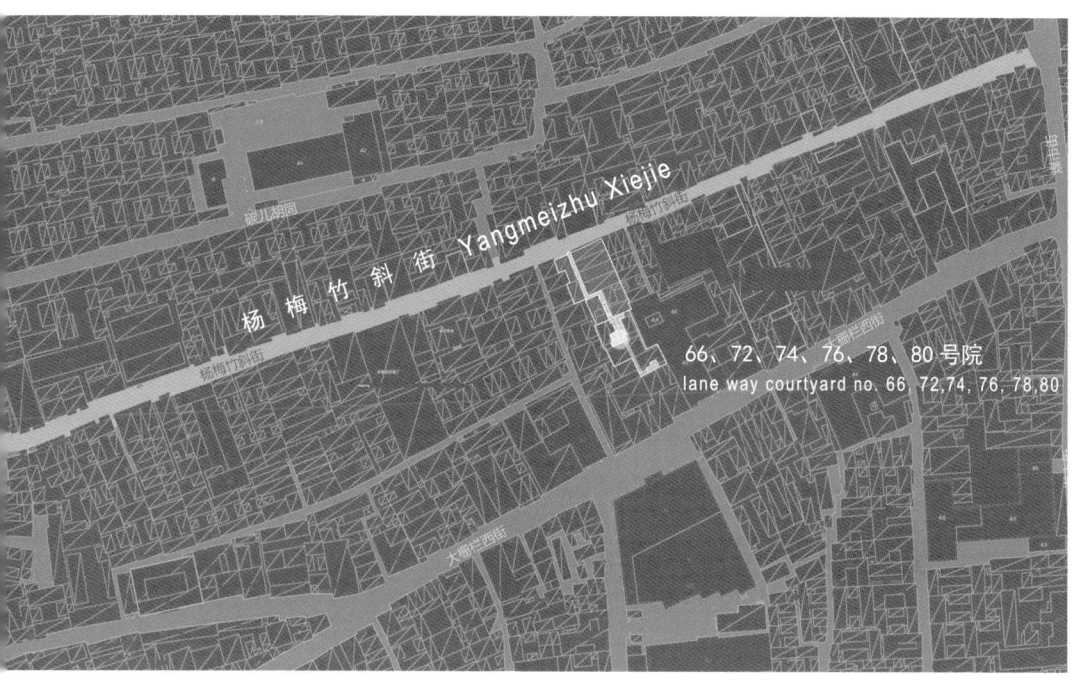

杨梅竹斜街 Yangmeizhu Xiejie

66、72、74、76、78、80号院
lane way courtyard no. 66、72、74、76、78、80

杨梅竹斜街区位
Yangmeizhu Xiejie zone

安住·杨梅竹斜街改造纪实与背后的思考

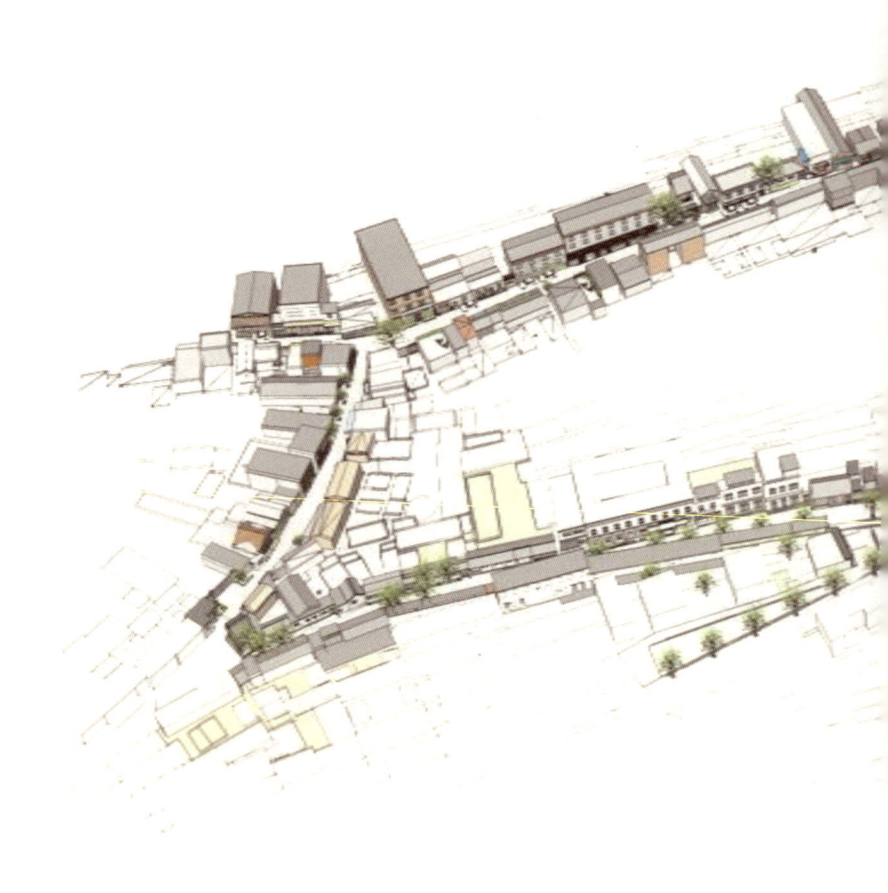

第一部分　方案实施

杨梅竹斜街鸟瞰效果图
Bird's eye rendered view of Yangmeizhu Xiejie

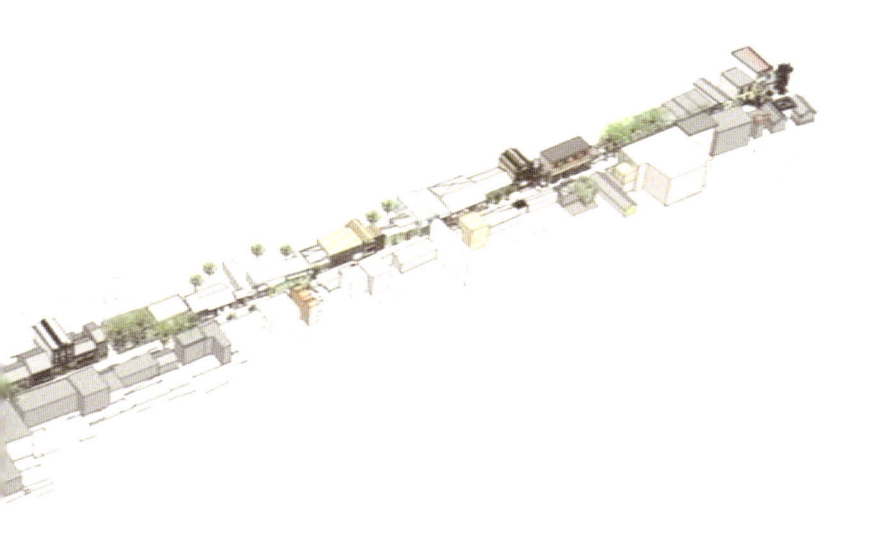

安住·杨梅竹斜街改造纪实与背后的思考

杨梅竹斜街南侧立面图

杨梅竹斜街北侧立面图

第一部分　方案实施

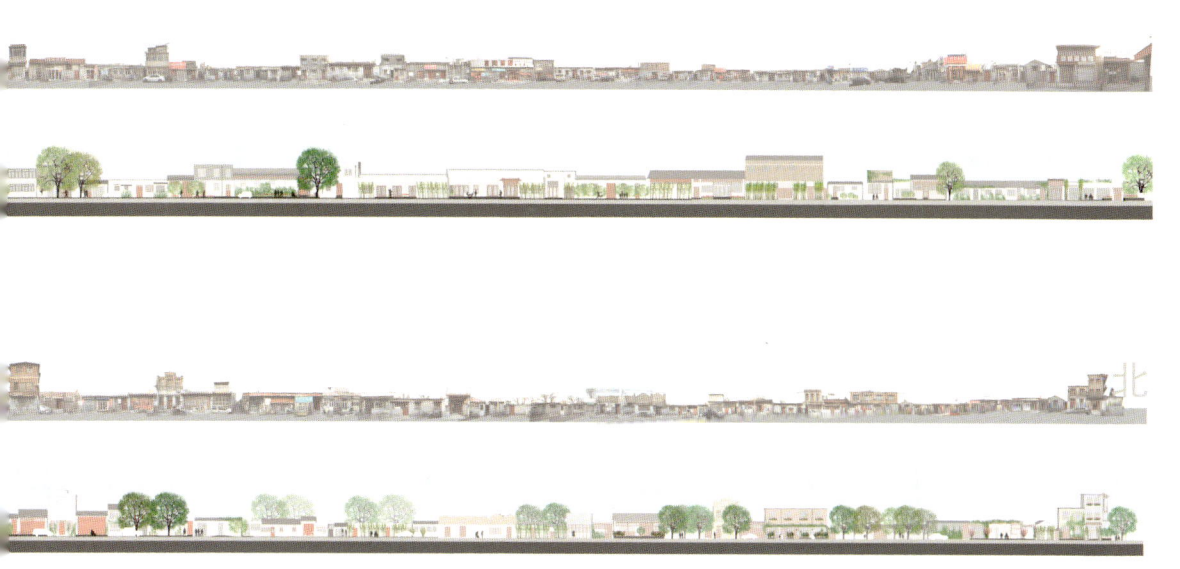

改造前后杨梅竹斜街街景对比（该图由无界景观 CUCD 与场域建筑合作绘制）
Comparative street view of Yangmeizhu Xiejie before and after renovations (image drawn jointly by View Unlimited CUCD and Approach Architecture Studio)

此外，这里坐落有东阁大学士梁诗正的宅邸，以及新中国成立前中共北平市委地下活动地点东升平澡堂，也留下了不同时代名人志士的足迹、故事，是一条著名的商业、文化和市井生活并存的街巷。

杨梅竹斜街改造始于 2009 年大栅栏地区更新计划，是政府主导、市场化运作模式的创新实践项目。采取微循环改造，保留老胡同的建筑肌理和市井文化，成为北京老城有机更新计划中的重要组成部分和示范项目。

2012 年 4 月，中国城市建设研究院无界景观工作室和场域建设承接了杨梅竹斜街环境更新项目。这是一个多层级协作项目，需要由政府、开发商、建筑师、景观设计师的协同合作，通过与每一户原住民和商户的沟通，来制定改造方案。在这一过程中，景观设计的任务是，厘清并保护胡同街区中固有的文化基因，并通过设计弥补其基因缺陷，激发它的活力。由于这个项目中存在特殊、复杂而交错的产权利益关系，参与的各方都需要随时调整方案。在重重矛盾中建立一种共同发展、可有机更新的模式，提供具有针对性的修缮方案，从而落实改善居民生活环境的设计目标，项目的每一步推进都是多方反复协商的结果。

第一部分　方案实施

改造前的杨梅竹斜街老建筑 (2012)
Old buildings of Yangmeizhu Xiejie before renovations

2012 年 8 月至 2013 年 12 月,项目实施过程中,设计师与每一户居民沟通,了解他们的切实需求,提出解决方案,在复杂的利益之间博弈,兼顾各方需求。最终,经过精心设计和反复论证,我们从街道立面、交通、绿化、铺装、市政设施、照明等 13 个方面对街道环境进行了有机更新。

在绿化方面,将拆违留下的空间处理为带有绿化的公共空间,通过花池、台阶的修砌将胡同公共生活与建筑联系起来,为胡同生活保留并扩展了舞台。我们鼓励居民通过自发绿化的方式使整条街道绿起来,从而培养了居民的公共意识。将拆违释放出来的空间置换成绿化的公共空间;用爬蔓植物覆盖暂时难以拆除的不雅建筑立面,保证街道景观的连续与统一;选择极富北京胡同特色的植物品种,延续老北京的胡同风情。

在铺装方面和基础设施方面,通过在铺装材料以及铺设方式上的创新,解决了普通烧结砖难以满足车行承重要求的难题,同时透气透水铺装材料的应用有效提高了胡同环境的舒适度,暑天不热蒸的街道得到老邻居们的赞许。铺装图案则利用算法技术进行设计,通过历史文化元素与新材料的结合,使得铺装有如一件记录时代信息的艺术品一般呈现在人们脚下,唤醒人们对于北京生活的触感与情感记忆。

第一部分　方案实施

改造前的杨梅竹斜街（2012.4）
Yangmeizhu Xiejie before renovations

安住 · 杨梅竹斜街改造纪实与背后的思考

Moreover, the street was also home to the residence of Liang Shizheng, Grand Secretary of the Eastern Cabinet, and the Dongshengping Bathhouse, a safe house and meeting place for the underground communists before the nation's liberation. Leaving behind the footprints and stories of famous figures and revolutionaries from different eras, it is a storied street in which business, culture, and the lives of ordinary people intersected. The renovation of Yangmeizhu Xiejie began with the Dashilan urban regeneration project in 2009, which employed innovative practices in government-led and market-oriented operations. By applying the principle of micro-intervention to preserve the architectural legacy and urban culture of the old hutongs, the renovation is a demonstrative project that has played a key role in the organic renewal of Beijing's old city.

In April 2012, View Unlimited Landscape Architects Studio took on the Yangmeizhu Xiejie renovation project. This was a multi-level and collaborative project that required significant coordination between the government, developers, architects, and landscape designers to formulate a renovation plan that involved extensive communication with the original residents and merchants. Our task during this process was to clarify and protect the inherent cultural DNA of the hutongs, while compensating for its defects by implementing a design that infused a greater sense of dynamism. Due to the existence of special, complex, and overlapping property rights and interests in this project, all parties involved were required to be flexible in adjusting their plans throughout the various stages of the project and in the face of various discrepancies. Considering this , although the design process was initiated by the client, through the efforts of the design team the hutong residents played an integral role.

From August 2012 to December 2013, the designers communicated with each of the residents during the implementation of the project to better understand their actual needs and proposed solutions. By considering these various requirements, the team strived to balance the complex interests of all parties involved and finally, after preparing a meticulous design that incorporated multiple rounds of proposals, was able to undertake the organic renovation of the hutong, which focused on 13 key aspects that included the street facades, transportation, greenery, paving, municipal facilities, lighting, and more.

In terms of greenery, the space created by demolishing illegal structures was transformed into green public spaces. Flower beds and steps were used to integrate the hutong residents' daily life with the buildings, preserving and expanding the hutong environment. To foster public awareness, we encouraged residents to practice spontaneous greening, thus replacing demolished illegal structures with green public spaces. Climbing plants were used to cover unsightly building facades that were difficult to remove to ensure a sense of continuity and unity throughout the landscape. Moreover, the team chose plant species with strong local characteristics to preserve the traditional charm of the hutongs.

In terms of paving and infrastructure, we used innovative paving materials and methods to provide a load-bearing solution for vehicles that ordinary sintered bricks could not offer. At the same time, the application of breathable and permeable paving materials enhanced the comfort of the hutong environment, allowing the alley to remain cool and comfortable in the summer heat, which was well-received by the elderly residents. The paving patterns used in the street's renovation were designed using algorithmic technology, which combined historical and cultural elements with new materials to create works of art that recorded information about the alley's history. Presented underfoot, each pattern awakens a host of tactile and emotional memories of Old Beijing .

改造前的杨梅竹斜街（2012.4）
Yangmeizhu Xiejie before renovations

第一部分　方案实施

改造后，杨梅竹斜街的基础设施得到改观 (2012.9)
After renovations, the infrastructure of Yangmeizhu Xiejie has been markedly improved

061

安住 · 杨梅竹斜街改造纪实与背后的思考

第 部分 方案实施

施工现场（2012.9）
At the construction site

安住·杨梅竹斜街改造纪实与背后的思考

改造后，杨梅竹斜街的基础设施得到改观 (2013.9)
After renovations, the infrastructure of Yangmeizhu Xiejie has been markedly improved

第一部分 方案实施

安住 · 杨梅竹斜街改造纪实与背后的思考

第一部分　方案实施

改造前（2012.4）与改造后（2013.9）
Before and after renovations

杨梅竹斜街的修缮与更新是一次寻找与本地居民及商户和谐共生的环境景观改造及保护发展模式的尝试。居民的生活环境将在这个过程中伴随着大家的共同努力逐渐得到改善，与此同时，通过鼓励居民参与到设计过程当中，间接地重塑了公共意识；通过新材料与历史材料的编织，原住居民的自由腾退与新型文化产业的入驻，当下主流文化与历史文化的整合，使得胡同文化在当代语境中，在时间上得以延续，空间上得以融汇，从而缩小近年来城市发展进程中带来的地区差异，也为区域未来经济的可持续发展提供舞台。

胡同的有机更新不是一个一蹴而就的过程，它处于复杂的利益博弈中，推进的过程需要不断地协商、反复。我们反对推倒重建式的改造，坚持微创与介入式的为胡同中的住户提供具有针对性的设计，协同多方，落实修缮与更新，希望通过这样的协同努力，杨梅竹斜街的有机更新可以以一种自然的、具体可感的姿态从街道本身中萌发出来。

从 2012 年到 2015 年，整个项目进展过程中，我们发现了原住民对于胡同改造的态度以及居民自发种植的行为，因此在后期工作中的角色发生了转变，从为居民设计到引导居民自发营造。

第一部分　方案实施

改造后的杨梅竹斜街 (2013.9)
Yangmeizhu Xiejie after renovations

The renovation and renewal of Yangmeizhu Xiejie is an attempt to implement a model of harmonious coexistence between the environment, landscape, and local residents and merchants. By harnessing the collective efforts of every stakeholder, the residents' living environment could gradually improve. At the same time, we are able to indirectly reshape public awareness for the project by encouraging residents to participate in the design process. Through the integration of new and historical materials, the voluntary vacating of the original residents, the occupation of new cultural industries, and a fusion of mainstream culture and historical legacy, the hutong culture can be perpetuated across eras and integrated spatially in a contemporary context, thus reducing the regional differences wrought by the urban development process in recent years and providing a platform for the sustainable economic development of the area moving forward.

The organic renewal of the hutongs is not an overnight process. Rather, it is a complex process involving various interests, which requires constant negotiation and iteration. We are strongly opposed to the "demolish and rebuild" approach and strive to offer targeted designs for hutong residents that employ a minimally invasive and interventionist approach. As such, we are dedicated to working together with all parties to implement the renovation and renewal of the hutongs, with the hope that through such collaborative efforts, the organic renewal of Yangmeizhu Xiejie can flourish in a natural and tangible way .

During the project's entire implementation process from 2012 to 2015, we discovered that the original residents embraced the transformation of the hutong and the concept of spontaneous planting. Thus, our role in the later stages of the project shifted away from designing for residents and focused more on how to guide residents to spontaneously create.

第一部分　方案实施

俯瞰改造后的杨梅竹斜街 (2014.9)
Overlooking the renovated Yangmeizhu Xiejie

安住 · 杨梅竹斜街改造纪实与背后的思考

113 号居民自建花园（2015.8）
Residents of No. 113 building their own gardens

072

第一部分　方案实施

三、胡同花草堂（2015.7—2015.9）
Hutong Flora Cottage
——杨梅竹斜街 66—76 号夹道社区营造计划
Public Space Renovation of 66-76 Yangmeizhu Xiejie Alley

杨梅竹斜街 66—76 号院夹道是一条最窄处仅 1 米、最宽处不超过 4 米的，全长 66 米的居民院落通道。经过前期的调研，2015 年的夏天，我们以志愿者身份参与了大栅栏领航员设计征集活动，对夹道进行了社区营造，我们将这一营造计划命名为"胡同花草堂"。

巷子中现居住了 5 户人家，1 棵老香椿树。其中 66 号和 74 号院的王氏家族最早于万历年间就定居于此，世代经营药铺。1949 年以后该院落经历了宿舍、书店、文具店等不同身份的变迁，如今借杨梅竹试点项目为契机，业态升级为咖啡馆和家庭博物馆，一定程度上成为杨梅竹斜街上老店铺历史变迁的一个缩影。夹道深处的 76 号原为杂院，部分住户搬迁，现由老北京魏氏一家四口居住，以及外来务工保安在此暂住。

第一部分　方案实施

中华人民共和国成立前这条巷子统称杨梅竹斜街 108 号，常见人口约 25 人。

66、74 号院：济安堂王家定居于此 30 余代，400 余年，是胡同里的老北京。前几年还有搬走的老邻居来重寻杨梅竹斜街 108 号，探访故人，重续旧缘。

72 号院：北房由公司租用，南房是东姐一家 3 口，居住 10 年，东姐热爱种植。

76 号院：前院由魏家 4 口居住，祖孙 3 代，居住 20 余年，爷爷爱种植。

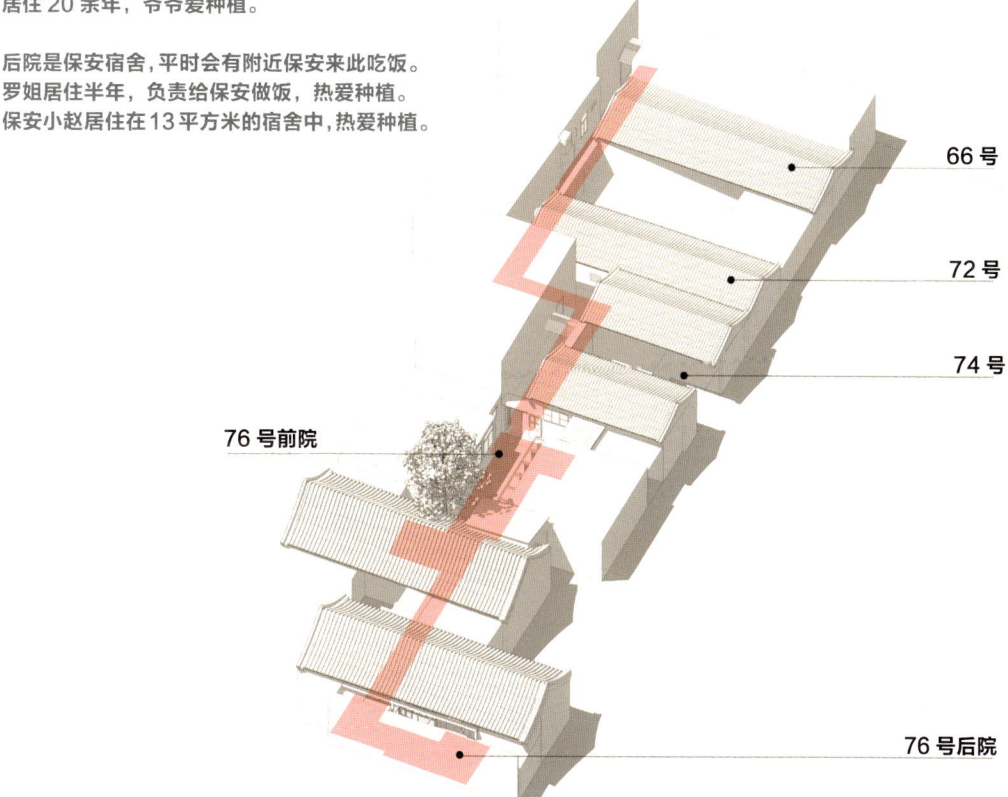

后院是保安宿舍，平时会有附近保安来此吃饭。罗姐居住半年，负责给保安做饭，热爱种植。保安小赵居住在 13 平方米的宿舍中，热爱种植。

66 号

72 号

74 号

76 号前院

76 号后院

调研结果显示"花草种植"是可以满足 25 位居民需求的"最大公约数"，这是我们设计构思"胡同花草堂"的依据。

The results show that "cultivating flowers and plants" is the "largest common denominator" in meeting the needs of the 25 residents, which is the basis for our design and the conception of the "Hutong Floral Cottage".

安住·杨梅竹斜街改造纪实与背后的思考

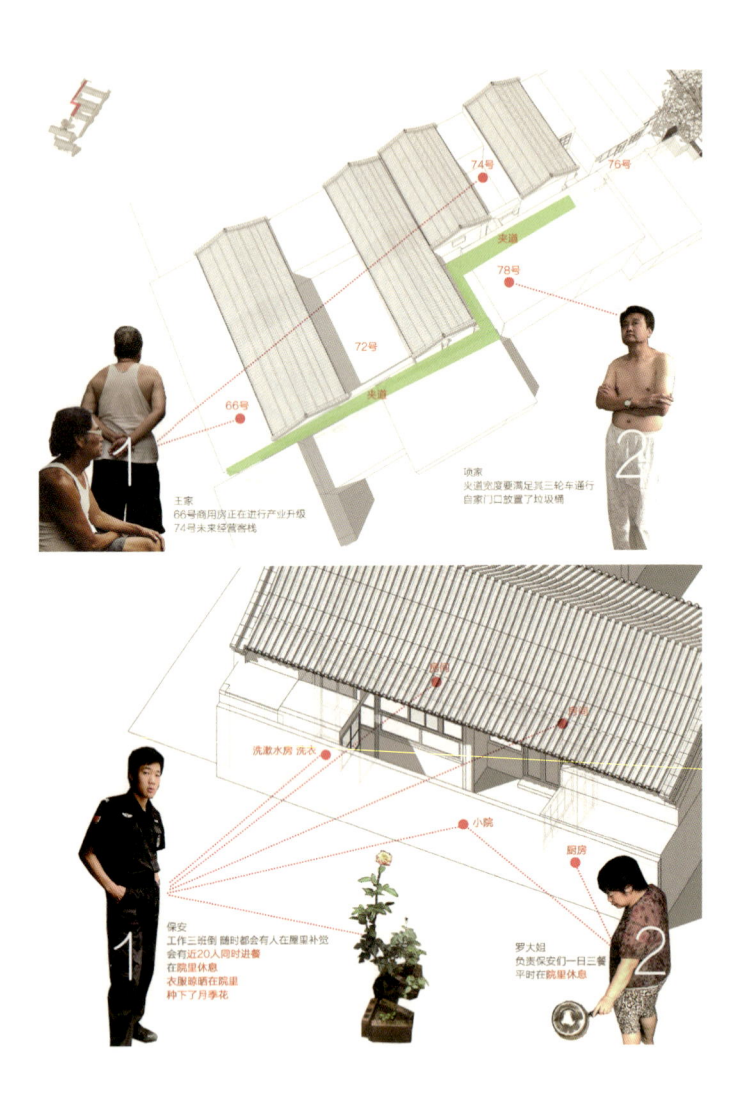

王家
66号商用店正在进行产业升级
74号未来经营客栈

项家
灭道宽度要满足其三轮车通行
自家门口放置了垃圾桶

保安
工作三班倒 随时都会有人在屋里补觉
会有近20人同时进餐
在院里休息
衣服晾晒在院里
种下了月季花

罗大姐
负责保安们一日三餐
平时在院里休息

076

第一部分　方案实施

房间　房间　房间

香椿树　小院

1　罗大姐
负责保安们一日三餐
大部分时间待在屋里
和魏家关系不错，邻里间互相帮忙
衣服都晾在香椿树外围的铁栅栏上

3　何安队长

魏家
祖孙3代 4口人
喜欢种花 春天会上树摘香椿
衣服晾晒在院里
部分杂物放在院里
老人腿脚不好 行动不便
喜欢在门口晒太阳
腿脚好的时候 在院里种菜 分给全筑街坊吃
女儿会在院里和小伙伴玩

2

077

安住·杨梅竹斜街改造纪实与背后的思考

"种植"是杨梅竹斜街 66—76 号夹道居民的共同爱好
"Planting" is the most commonly shared interest of residents from the alley of Nos. 66-76 Yangmeizhu Xiejie

由此可见，这条夹道由世代居住的老北京原住民、外来务工人员等多种类型的人群共同使用。不同的身份背景使他们各自的需求和生活方式也各不相同，我们通过与杂院居民的沟通，了解他们的生活方式，最终决定以建立"胡同花草堂"的方式介入，借用居民自己的智慧改造现有的夹道和生活空间，在有限的条件下提高杂院生活质量，引导新的生活模式。通过"绿植花草"新经验的分享等活动，为不同人群提供交流的可能性，减少矛盾的产生，也让更多的人们能够了解胡同日常生活的情境。

夹道狭长曲折，在杨梅竹斜街杂院空间中具有唯一性。但也存在空间单调乏味，缺乏变化的问题。经过前期调查，设计师们针对夹道地面不平整、墙面污损、缺乏收纳空间、排水不畅等问题重新梳理规划了夹道空间，在排水、绿化、基础设施等方面进行了建设，增加了三处雨棚，局部拓展了院落公共区域，形成节奏不断变化，富有趣味的夹道杂院空间。

在排水方面，修整建筑散水用以排除雨水，保护墙基免受雨水侵蚀，将现状散水修补整齐，不牺牲胡同宽度，保证通行顺畅，占用最小的宽度，摆放植物以美化环境。

The alley between Nos. 66-76 Yangmeizhu Xiejie is a narrow passageway featuring a cluster of residential courtyards that is only 1 meter at its narrowest and no wider than 4 meters at any one point, with a total length of 66 meters. After conducting preliminary research, we volunteered to participate in the Dashilan Pilot Call for Proposal in the summer of 2015 and implemented a community building initiative for the alley, which we named the "Hutong Flora Cottage" project.

The alley is composed of 5 households and 1 old Chinese scholar tree. Among them, the Wang family who live in No. 66 and No. 74 settled here as early as the Wanli period and operated a pharmacy for generations. After 1949, the cluster of courtyards experienced drastic changes in its identity, which included the introduction of dormitories, bookstores, and stationery stores. Today, seeing the potential of the Yangmeizhu pilot project, local businesses have expanded to include a coffee shop and family-themed museum, which represents a microcosm of the historical changes to the old shops on Yangmeizhu Xiejie. No. 76 in the depths of the alley was originally a mixed-use courtyard. However, after some of the residents moved out, it became inhabited by not only the Wei family, an Old Beijing family of four, but also migrant workers who worked as security guards. This change reflected the alley as a whole, which was shared by various people groups, including both Old Beijing locals who had lived here for generations and transient migrant workers. Due to these differing backgrounds, the residents had a wide range of needs and led different lifestyles, which we were able to learn more about by communicating with them . Ultimately, we decided to involve ourselves by establishing the "Hutong Flora Cottage", which leveraged the wisdom of residents to transform the existing alley and living environment, improve the overall quality of life in the courtyard cluster under constrained conditions, and introduce a new way of life. By offering a shared experience through the act of planting, we could provide a platform for communication between these different people groups to reduce potential conflicts and allow a broader range of people to understand the daily life in the hutongs.

The alley is a long and winding space that is unique among the Yangmeizhu Xiejie hutongs, however the space suffered from a sense of monotony. After conducting a preliminary investigation, the design team re-imagined the alley and addressed various issues such as the uneven paving, dirty walls, lack of storage space, and poor drainage that afflicted the area. After this, the team carried out renovations to improve the alley's drainage, greenery, and local infrastructure by introducing three rain shelters, expanding the public areas within the courtyard cluster, and creating an interesting and ever-evolving cadence that accompanied the laneway and courtyards.

In terms of drainage, the buildings' runoff systems were modified to remove rainwater more effectively and protect the foundation from further erosion. Moreover, the team focused on restoring the existing system without sacrificing the width of the hutong to ensure smooth passage and introduced additional greenery to beautify the environment.

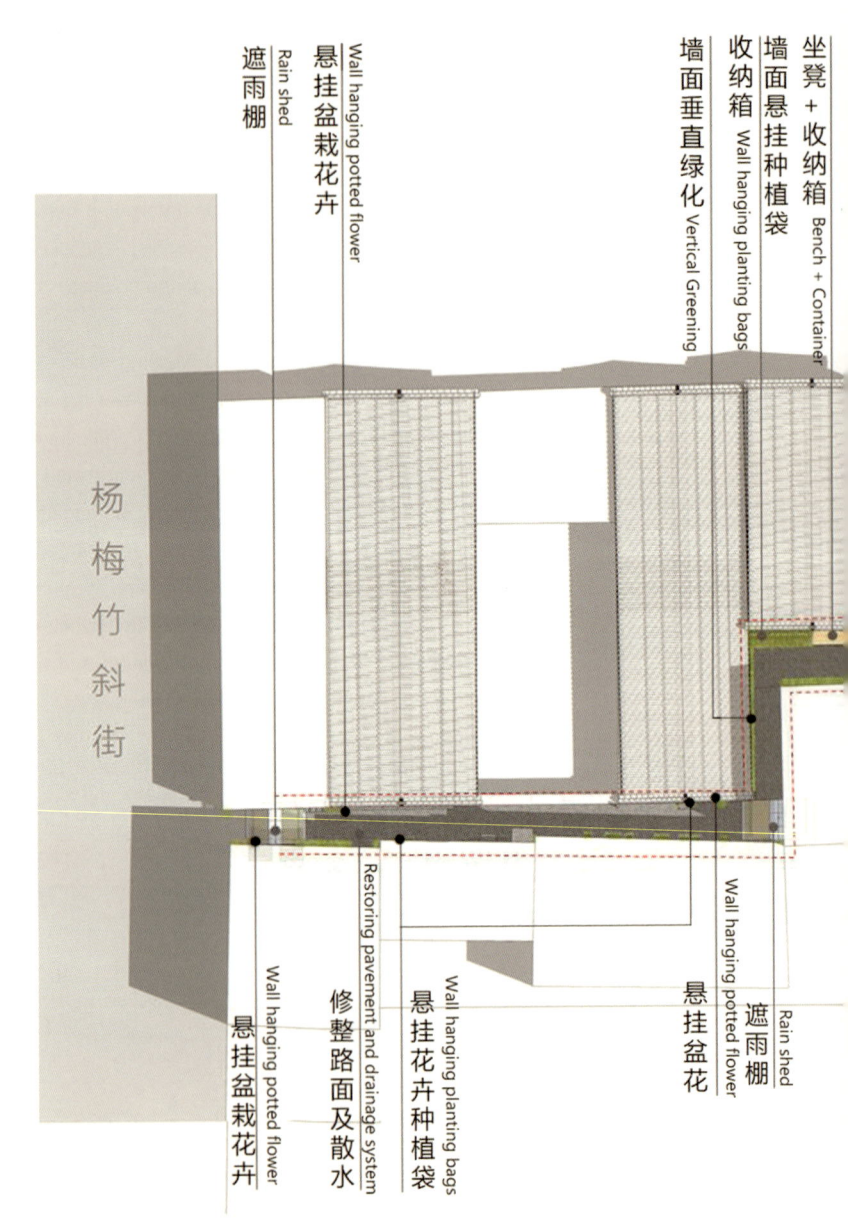

66—76 号杂院夹道平面图

Floor plan for the courtyard of Nos. 66-76

坐凳 + 收纳箱 Bench + Container
墙面悬挂种植袋 Wall hanging planting bags
收纳箱
墙面垂直绿化 Vertical Greening
悬挂盆栽花卉 Wall hanging potted flower
遮雨棚 Rain shed
杨梅竹斜街
修整路面及散水 Restoring pavement and drainage system
悬挂花卉种植袋 Wall hanging planting bags
悬挂盆花
遮雨棚 Rain shed
悬挂盆栽花卉 Wall hanging potted flower

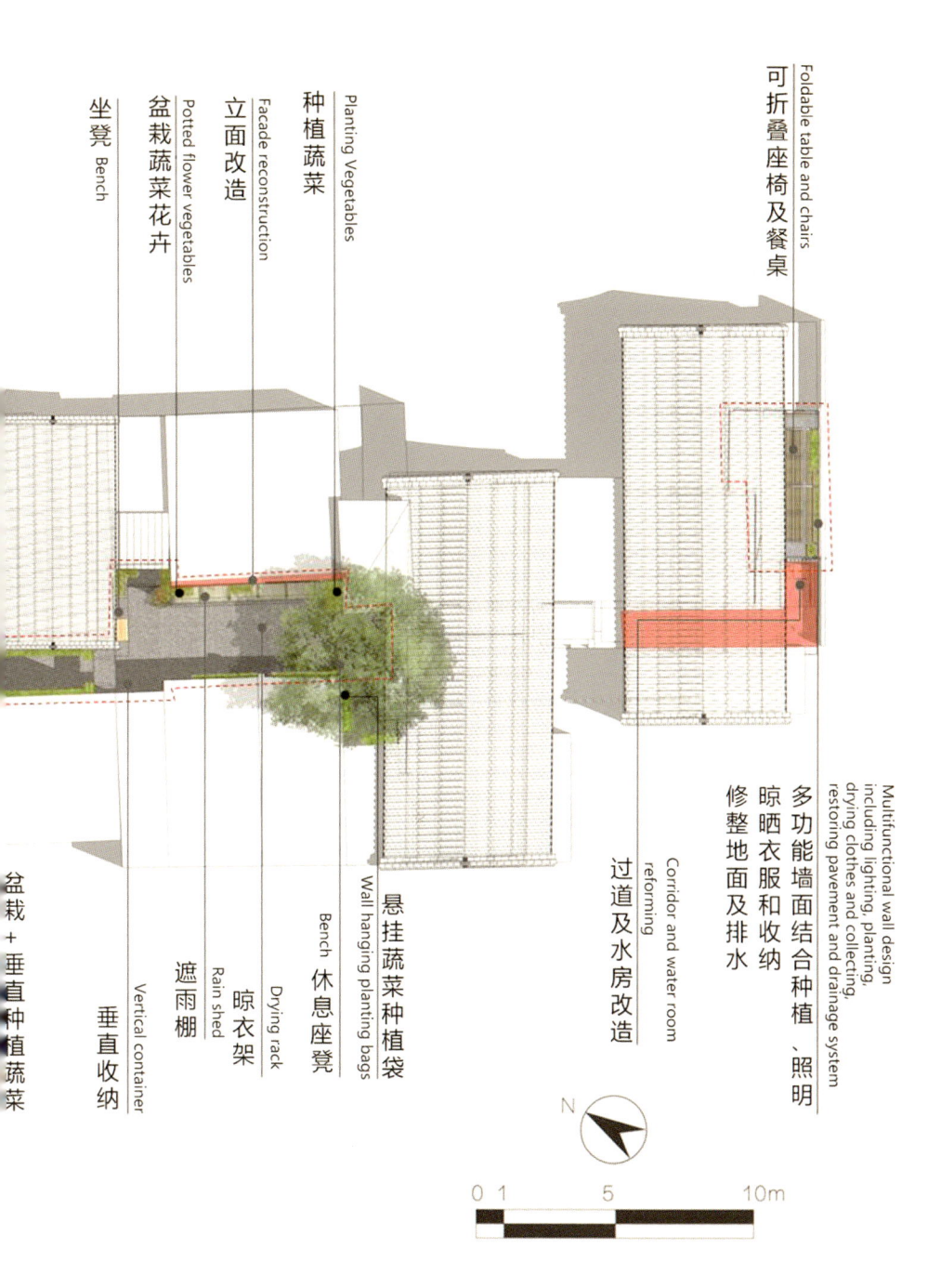

第一部分　方案实施

可折叠座椅及餐桌 Foldable table and chairs

种植蔬菜 Planting Vegetables

立面改造 Facade reconstruction

盆栽蔬菜花卉 Potted flower vegetables

坐凳 Bench

多功能墙面结合种植、照明 晾晒衣服和收纳 修整地面及排水 Multifunctional wall design including lighting, planting, drying clothes and collecting, restoring pavement and drainage system

过道及水房改造 Corridor and water room reforming

悬挂蔬菜种植袋 Wall hanging planting bags

休息座凳 Bench

晾衣架 Drying rack

遮雨棚 Rain shed

垂直收纳 Vertical container

盆栽 + 垂直种植蔬菜

N

0　1　　5　　10m

083

在绿化方面，见"缝"插绿，增加胡同绿量。有效利用有限的种植空间，采用盆栽与垂直绿化相结合的方式，不破坏建筑墙面。观赏与实用相结合。选择居民喜爱的具有食用和观赏功能的植物。

在基础设施方面，收纳电线，让杂乱的墙上视觉空间变得有序；增加扶手，让老人在胡同中行走更加便捷、安全。

建立以花草为媒的"胡同花草堂"，通过种花种草这种大家共同的爱好，让大杂院里的不同人群有了共同话题，缓解逼仄的生活空间带来的生活压力。鼓励每个胡同居民通过自家花草种植美化胡同环境，分享有趣味的种植经验和健康理念。这种新的种植和果实分享经验是生活环境得到改善后所带来的对生活方式的一次升级。

改造后效果图
post-renovation renders

In terms of greening, plants were introduced to increase the amount of greenery in the hutong. Although limited, these planting spaces were effectively utilized by combining potted plants with vertical greening systems that preserved the integrity of the building's walls. Moreover, the design team focused on fusing decorative and practical elements by selecting plants preferred by local residents that were ornamental yet edible.

In terms of infrastructure, the electric wires that haphazardly adorned the walls were organized and stowed away to maintain a more orderly appearance, while handrails were added to make it easier and safer for the elderly to walk through the hutong.

The "Hutong Flora Cottage" was established as a medium for planting flowers and plants, which allowed for courtyard residents from different backgrounds to bridge a common topic through a shared hobby, thus relieving the pressure of living in a cramped living space. In addition, each resident was encouraged to plant flowers and plants in their own homes to beautify the hutong environment, partake in the community experience, and embrace a healthier lifestyle. Through this creative idea to plant and share fruit and vegetables, the residents' surrounding environment and quality of life were tangibly improved.

安住·杨梅竹斜街改造纪实与背后的思考

第一部分　方案实施

增加种植和桌椅收纳设施，解决 20 位保安每天没有地方坐下吃饭的难题，让暂住北京的人们也能留下美好的记忆。

additional planting facilities and storage installations that work as tables and seatings, so that the 20 securities guards would have some testing places to eat everyday. Temporary residents in Beijing would also have a good memory here.

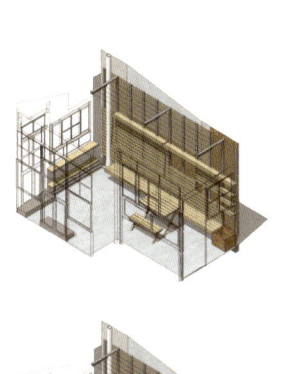

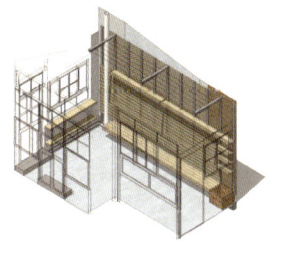

夹道改造重点分析及改造后效果图
Diagram with key points analysis of alley renovation and post-renovation renders

087

安住·杨梅竹斜街改造纪实与背后的思考

夹道改造施工过程中（2016）
During the alley renovations

第一部分　方案实施

四、杨梅竹斜街社区治理与公共空间提升（2017—2018）
Yangmeizhu Xiejie Community Governance and Public Space Enhancement

2017年10月，随着北京城区腾退工作的加速进展、老城人居环境标准的不断提升，杨梅竹斜街因其特殊的历史人文价值，成为东、西城十大公共空间改造试点之一。2018年，杨梅竹斜街社区治理与公共空间提升项目启动。无界景观作为景观设计方承接了该项目，提出了包括提升整体公共空间质量、完善绿化种植系统、城市照明系统、健身系统、智慧人本空间系统、停车管理系统和垃圾收集分类系统等方面的系统性的解决方案，并通过文化系统梳理和重要节点设计，结合智慧人本空间营造展现更深层次的文化内涵。延续"胡同花草堂"的共享共建理念，尊重、支持和鼓励居民自发美化街区环境，激发社区环境持久的生命力和活力，促进邻里交往，提升居民幸福感和归属感。

回顾6年的改造实践，居民自发行为所留下的、设计方所抛弃的，以及在与居民沟通、互动中日渐形成的"设计"启发了我们，使我们在持续的设计实践中始终遵循着"设计逻辑不能违背居民日常生活逻辑"的理念。"设计"将与居民的日常生活共同成长，以未完成的形态存在并期待其继续发展。

第一部分　方案实施

In October 2017, with the accelerated redevelopment of Beijing's urban areas and continuous improvements in the living standards of the old city, Yangmeizhu Xiejie became one of the top ten pilot projects for the renovation of public spaces in Dongcheng and Xicheng districts because of its special historical and cultural value. Subsequently, in 2018 the Yangmeizhu Xiejie Community Governance and Public Space Enhancement Project was launched. As the landscape design company undertaking this project, View Unlimited proposed a systematic solution that focused on improvements in the overall quality of public space, green planting, urban lighting, fitness facilities, smart city infrastructure, parking management, and garbage collection and classification. By ascertaining important cultural nodes and integrating them into a smart and digitally-driven environment, the project showcased the deep-rooted cultural connotations of the old city.

Looking back on the past six years, the continued concept of the "Hutong Flora Cottage" has left us inspired by the perceptible results of residents' spontaneous actions, the elements that were archived, and a "design" that has continued to evolve through communication and interaction with local residents. Throughout our approach, we have maintained that "the logic of design should never contradict the logic of residents' daily lives". A design should grow together and intersect organically with residents' daily lives. Existing in an unfinished form, we can only anticipate its continued development. As such, we support and encourage residents to spontaneously beautify their neighborhood environment, to infuse an enduring vibrancy and dynamism into their community, promote greater communication, and enhance each resident's sense of happiness and belonging.

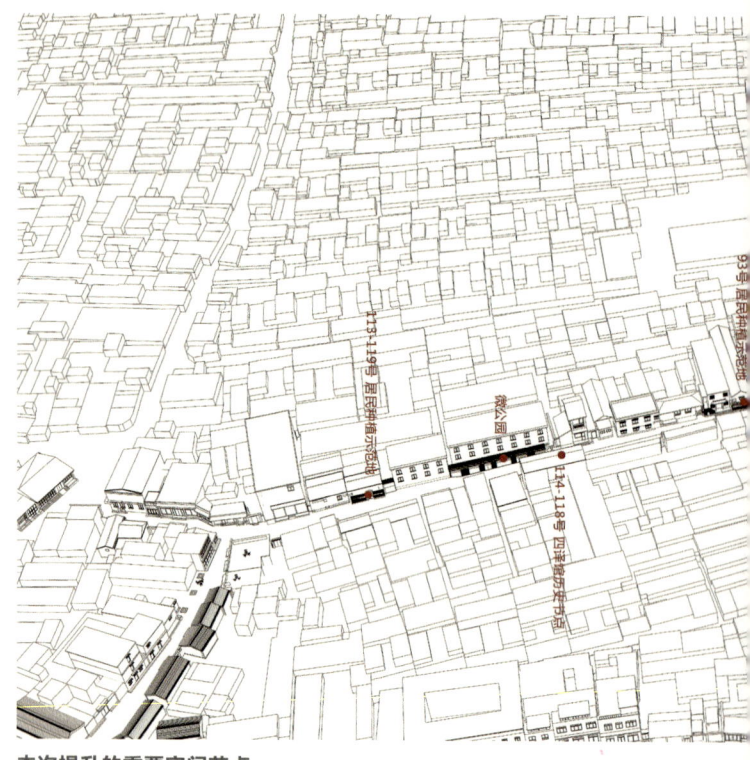

本次提升的重要空间节点
Important spaces during this round of improvements

第一部分　方案实施

75号 世界书局历史节点
90号 书局印刷厂历史节点
66号 安馨历史节点
61号 资西会馆历史节点
76号 胡同花草堂
49号 新儿童书局历史节点
44号 居民种植示范地
16号 梁诗正故居历史节点
25号 和合会馆历史节点
6号 东升平浴池历史节点
9号 居民种植示范地
4号 泰丰楼历史节点

093

2015 年、2018 年 76 号院实景照片
Site Photos of courtyard 76 from 2015 and 2018

第一部分　方案实施

改造后的 76 号院实景照片（2023）
Site Photos of courtyard 76 after renovations（2023）

安住 · 杨梅竹斜街改造纪实与背后的思考

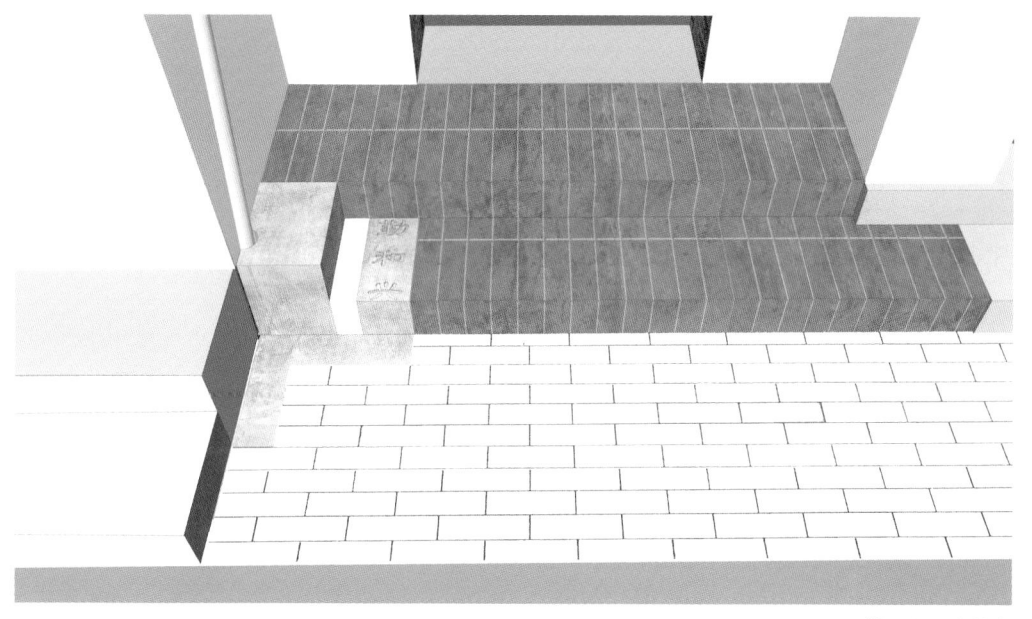

第一部分　方案实施

勤和兴历史节点
Qinhe xing historical node

拆除现有破损台阶和铺装，新建台阶和铺装，以铺装记录此节点历史信息。
The existing damaged steps and pavement were removed, and new steps and pavement were installed so as to record the historical information of this node.

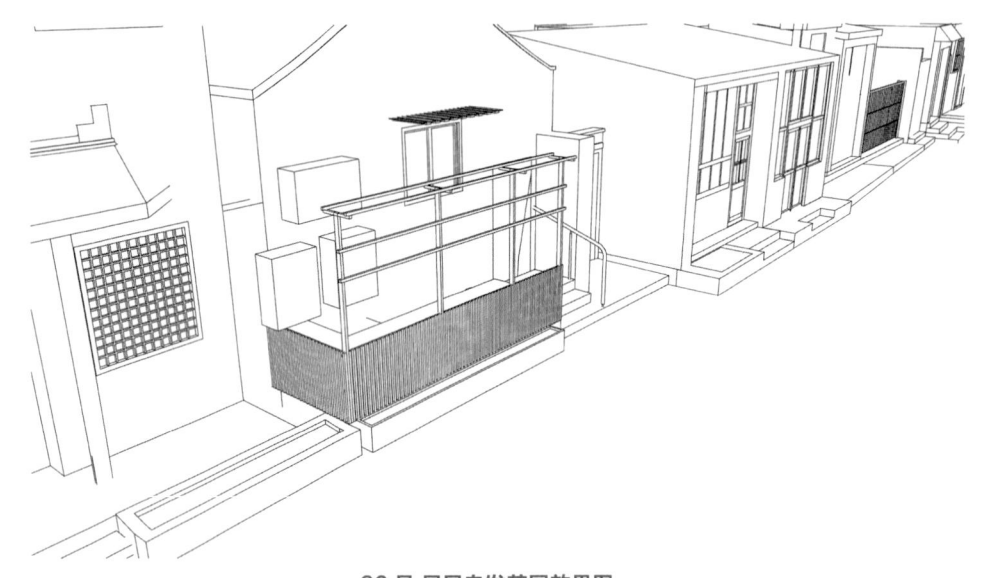

93 号 居民自发花园效果图
Impression drawing of spontaneous gardening by residents at no. 93

第一部分　方案实施

悬挂于墙面上的是胡同微公园 2012 年改造前的照片（2014）
A photo of the Hutong Mini Park before the 2012 renovation is displayed on the wall (2014)

第一部分　方案实施

2012 年改造后的胡同微公园，地面高差设计解决了机动车停车占地问题，
居民可以晾晒被子，日常健身，参与节假日活动（2014）

Following the 2012 renovation, the Hutong Mini Park features a design that addresses the issue of
motor vehicle parking occupancy by utilizing variations in ground height. Residents can continue to air their quilts,
engage in daily fitness activities, and participate in holiday events (2014)

安住·杨梅竹斜街改造纪实与背后的思考

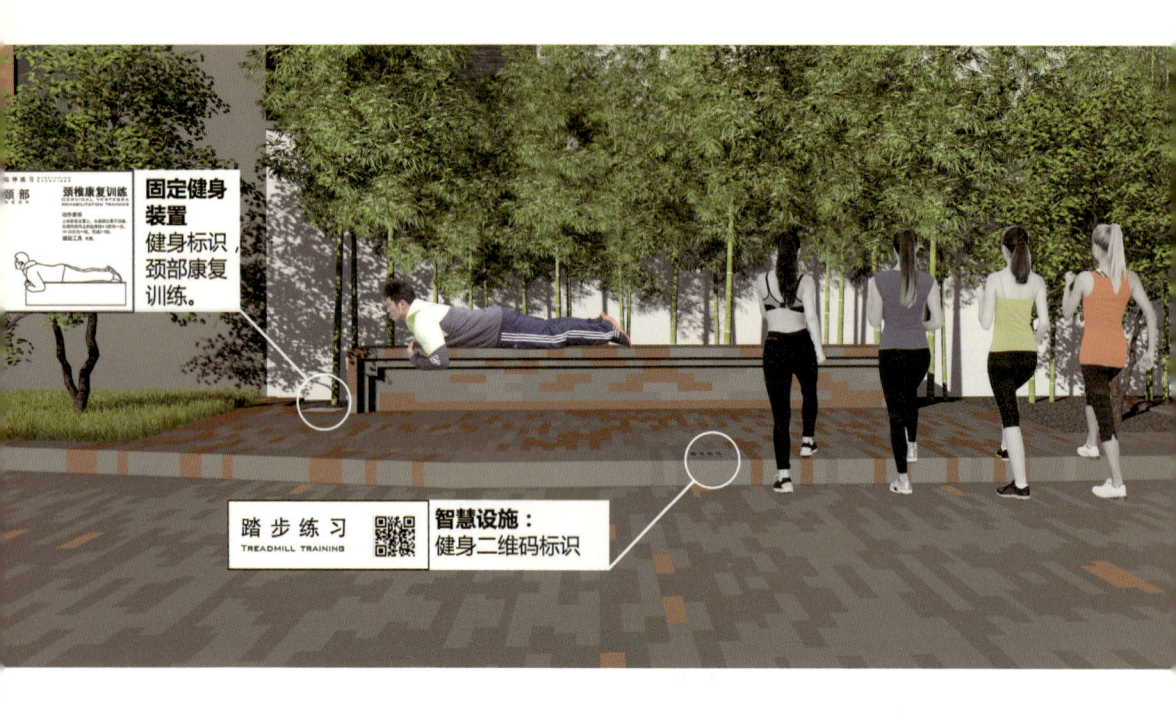

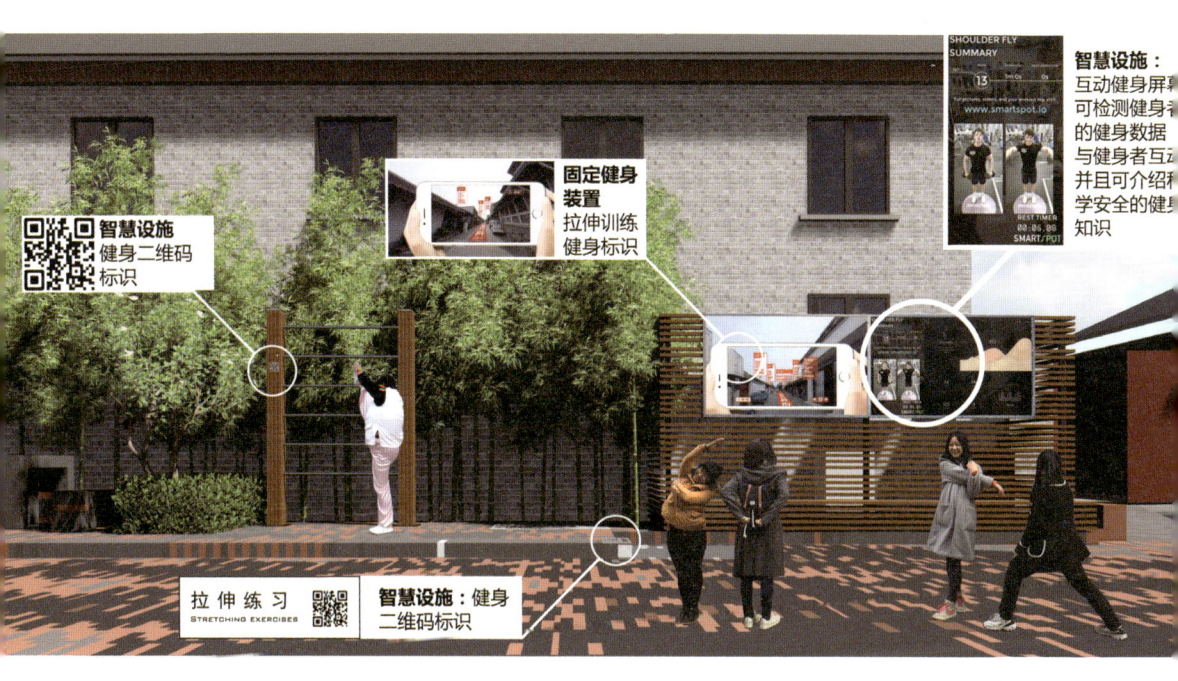

第一部分　方案实施

智慧设施：
互动健身屏幕可检测健身者的健身数据，与健身者互动并且可介绍科学安全的健身知识

智慧设施：
健身二维码标识

固定健身装置
拉伸训练
健身标识

拉伸练习
STRETCHING EXERCISES

智慧设施：健身二维码标识

2018 年，与智慧城市设施结合的胡同微公园改造效果图（上图）
改造后，杨梅竹斜街的居民日常健身活动（下图）
An impression drawing of a Hutong Mini Park Fitness Centre combined with Smart City facilities in 2018(above)
Residents of Yangmeizhu Xiejie engage in daily fitness activities after the renovation(below)

103

安住·杨梅竹斜街改造纪实与背后的思考

第一部分　方案实施

在设计团队的引导下，居民开始自发改善屋顶环境。2018 年夏，邻居们在 66 号屋顶花园参加活动（齐欣摄）
Under the guidance of the design team, residents spontaneously began to improve their rooftop environments. During the summer of 2018, neighbors attend activities on the rooftop garden (Photo by Qi Xin)

105

五、一尺大街改造设计及实施情况（2012—）
Renovation and Transformation of Yichi Street

"一尺大街"属于杨梅竹斜街西段，名称源于清代学者陈宗蕃 1931 年的著作《燕都丛考》，已有近百年历史，为了便于管理，于 1965 年并入杨梅竹斜街。它是北京最短的胡同，长约 25 米。历史上，此街曾经容纳 6 家店铺，经营刻字、酒馆、铁匠和理发等业务，展现了老胡同的商业风貌与手工艺传统，是老北京胡同的缩影，也是研究和体验老北京生活的有趣场所。

Yichi Street, situated in the western stretch of Yangmeizhu Xiejie, takes its name from the 1931 work "Yandu Congkao" by Qing Dynasty scholar Chen Zongfan. With nearly a century of history, this alley was integrated into Yangmeizhu Xiejie in 1965 to streamline management. Remarkably, it holds the distinction of being the shortest hutong in Beijing, measuring roughly 25 meters in length. Historically, Yichi Alley was home to six establishments offering services such as engraving, blacksmithing, barbering, and a tavern. These businesses reflected the commercial and artisanal essence of traditional hutongs, rendering Yichi Alley a microcosm of old Beijing life—a captivating locale to immerse oneself in the city's rich historical allure.

第一部分　方案实施

嵌入"一尺大街"两端地面的标识牌（2018）
Street name signs embedded at both ends of Yichi Street（2018）

在 2012 年杨梅竹斜街环境更新项目中，为了强化其历史文化的地标角色，设计团队提出了恢复"一尺大街"原名的建议，但未获政府机构采纳，街道改造设计方案也未能实施，最终，仅在地面上嵌入了"一尺大街"字样的标牌。

在 2018 设计团队再次提出了对"一尺大街"的更新改造方案。该方案以艺术手法进行地面铺装设计，将汉白玉、青石等传统材料与花岗岩交织拼接。汉白玉与青石是北方长久以来最能象征城市营造的物质文化意向，在此通过抽象与写意的地面铺装，唤醒对于历史文化脉络的追忆，为胡同增添了一种历史层叠的视觉效果。地面铺装与入户台阶衔接处的石材表面镌刻原 6 家店铺的相关历史信息，以不同字体呈现。隐藏在花池和座凳内的声音装置播放一段段真实录制的老北京商贩叫卖声，声音的出现与消失，就像记忆中的片段。设计旨在汇聚个体记忆，激发集体共鸣，促使人们去探索更多老北京的故事。

第一部分　方案实施

As part of the 2012 Yangmeizhu Xiejie environmental renewal project, the design team suggested restoring the original name of "Yichi Street" to enhance its historical significance. However, this proposal wasn't approved by the government. In the end, only a plaque with the name "Yichi Street" was installed on the ground, without any further renovation work.

In the 2018, the design team revisited plans for revitalizing Yichi Street. Their approach involved artisanal ground paving using a blend of traditional materials like Han white jade, bluestone, and granite. In particular, Han white jade and bluestone have long symbolized urban heritage in the north. Through abstract and evocative ground patterns, the project aimed to reconnect people with the area's historical and cultural roots, adding a visually rich layer to the alley. Additionally, historical information about the original six shops was engraved in different fonts on the surfaces where the ground paving meets the doorstep. Hidden within the flower pond and stools, a sound installation plays recorded voices of real old Beijing street vendors. The emergence and disappearance of these sounds, like fragments of memory. The design aims to gather individual memories, evoke collective resonance, and encourage people to explore more stories of old Beijing.

安住 · 杨梅竹斜街改造纪实与背后的思考

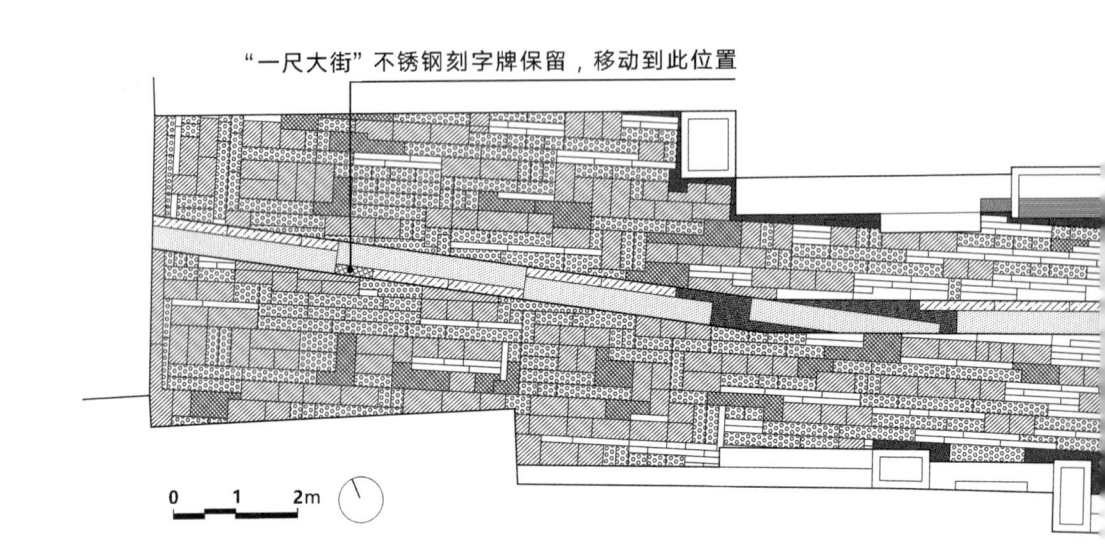

"一尺大街"不锈钢刻字牌保留，移动到此位置

"一尺大街"铺装设计平面图
Layout plan for the paving design of Yichi Street

"一尺大街"不锈钢刻字牌保留，移动到此位置

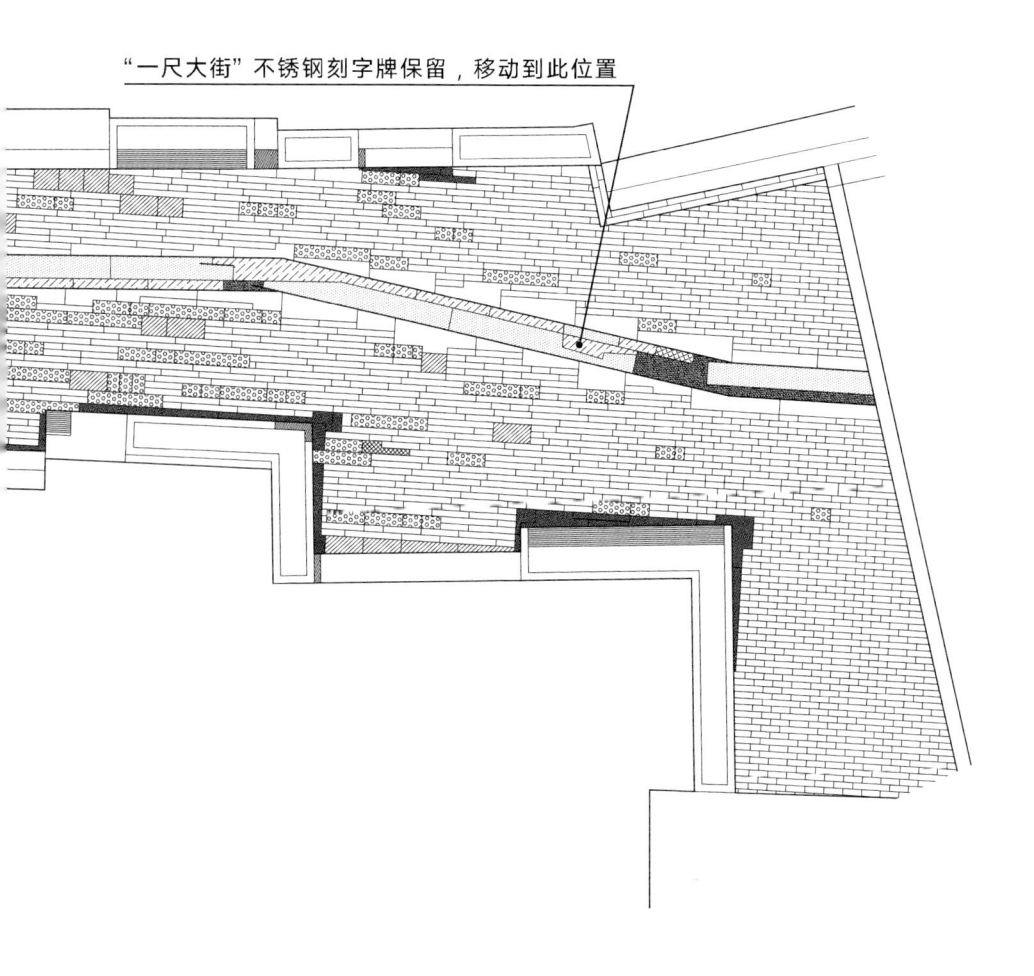

"一尺大街"衔接了杨梅竹斜街与东琉璃厂街，构建了一个连续而完整的步行体验空间，成为承载地方文化基因的重要载体，也成为人们心中的记忆锚点。然而遗憾的是，由于多种原因，这一设计方案至今未能付诸实施。

今天，"一尺大街"发生了新的变化。这一转变源自杨梅竹斜街 166 号居民郎国丽女士。她在 2016 年胡同花草堂种植展中受到启发，于自家小院开辟出一片名为"一尺花屋"的花草天地。两年后的 2018 年，无界团队邀请她参与花草堂种植展，向市民和游客展示了她的园艺才华。2020 年至 2023 年的疫情时期，郎国丽女士带动邻里（杨梅竹斜街 168 号）共同参与。她们不仅在自家门前种花植草，还将这份绿意和愉悦延伸至街头巷尾。

第一部分　方案实施

Yichi Street links Yangmeizhu Xiejie with the East Liulichang Street, forming a seamless pedestrian experience. It has evolved into a significant hub for preserving local cultural roots and serves as a cherished memory marker for many. However, regrettably, for various reasons, this design concept remains unimplemented to this day.

Today, Yichi Street has undergone a new transformation, all thanks to Ms. Lang Guoli, a resident of 166 Yangmeizhu Xiejie. Inspired by the 2016 Flora Cottage exhibition, she created a charming flower garden called "Yichi Flower House" in her courtyard. Two years later, in 2018, she was invited by the View Unlimited Team to participate in the annual flower and plant exhibition, showcasing her gardening talents to locals and tourists alike. During the COVID-19 pandemic from 2020 to 2023, Ms. Lang Guoli encouraged her neighbors at 168 Yangmeizhu Xiejie to join in. Together, they not only planted flowers and greenery in front of their homes but also spread this greenery and joy throughout the streets and alleys.

郎国丽女士与设计师切磋种植经验
Ms. Lang Guoli was exchanging planting experiences with the designer

第一部分　方案实施

2023年杨梅竹斜街种植大赛作品评选期间，郎国丽女士（左二）与家人和设计师合影
During the 2023 Yangmeizhu Xiejie planting contest , Ms. Lang Guoli (second from the left) posing with family and designers

安住·杨梅竹斜街改造纪实与背后的思考

2018 年北京国际设计周期间杨梅竹斜街街景（摄影：齐欣）
Street view of Yangmeizhu Xiejie during the 2018 Beijing Design Week (Photo by Qixin)

第二部分
展览传播

Part 2 Exhibition Communication

Arte
Architettura
Cinema
Danza
Musica
Teatro
Archivio Storico
Biennale
2016
Carnevale 30.01-7.02
Architettura 28.05-27.11
Danza 17.06-26.06
Teatro 27.07-14.08
Cinema 31.08-10.09
Musica 7.10-16.10
Attività permanenti

一、安住·平民花园
HOME　Communal Garden

——2016年第15届威尼斯国际建筑双年展中国馆作品
The 15th Venice International Architecture Biennale China Pavilion,2016

安住·杨梅竹斜街改造纪实与背后的思考

第 15 届威尼斯国际建筑双年展中国馆场地 google 卫星图
Google satellite map of the China Pavilion site
for the 15th Venice International Architecture Biennale

120

第二部分　展览传播

第 15 届威尼斯国际建筑双年展于 2016 年 5 月底正式开幕，展览持续 6 个月。本届建筑双年展的总策展人是 2016 年普里兹克建筑奖获得者亚力杭德罗·阿拉维纳。主题是："来自前线的报道"。

The 15th International Architecture Exhibition kicked off in late May 2016 and ran for six months. Curated by Alejandro Aravena, the 2016 Pritzker Architecture Prize laureate, the theme was "Reporting from the Front".

安住·杨梅竹斜街改造纪实与背后的思考

2016年，无界景观团队的杨梅竹斜街66—76号夹道营造作品有幸作为中国国家馆的参展作品之一入选了第15届威尼斯国际双年展。

威尼斯建筑双年展于1895年首次举行，具有百年的历史。它所包含的国际视觉艺术双年展与建筑双年展分单、双年轮流举行，它们与德国卡塞尔文献展、瑞士巴塞尔艺术博览会并驾齐驱，成为世界建筑艺术和学术界最具影响力的盛事。

中国国家馆的参展方案经过文化部评审委员严格、认真地评选，最终确定了策展人梁井宇的方案。梁井宇携手当代中国若干组优秀的建筑、设计和艺术团队，共同构建了主题为"平民设计，日用即道"的中国国家馆，以契合今年建筑双年展总策展人，普利策奖获得者建筑师亚力杭德罗·阿拉维纳提出的"来自前线的报道"（Reporting from the Front）这一主题。

作为第15届中国国家馆参展的九家机构之一，我们为中国馆带去了曾在2015设计周大栅栏设计社区上展示的公共环境设计试点"安住·平民花园"（更生·相续——杨梅竹斜街夹道杂院改造）项目，该项目分为两个部分展出。室内展览内容为多媒体视频，展现我们近年来在老旧街区改造计划中采集整理的杨梅竹斜街的日常街景，以及该街66—76号院夹道居民的衣食住行；展示我们对社区营造的思考与解决方案以及胡同花草堂的产生过程。

室外装置"平民花园"是用北京老旧街区中随处可见的废弃物组合而成的花池，意在反映老旧街区居民们的生存状态：无奈＋希望。装置是我们依据杨梅竹斜街66—76号院夹道的平面图，用胡同居民的日用废弃物搭建而成的"实体"夹道。这些关乎人们日用的弃物无不承载过美好的心念，读取这些弃物中储存的信息使我们与之产生了共振，并以这种方式讲述夹道里的故事。在这组装置中我们设置了五组视频播放器，视频内容是设计师们开展前一周录制的、没有经过任何剪辑的夹道中五户居民的生活片段以及"花草堂"实景。同时，展览现场还有两组视频是通过网络实时转播的。参观者可以通过视频实时看到夹道中的景象，而远在北京的居民们也可以通过视频看到展览现场的实况（包括中国馆的开幕式）。

这是一个在6个月的展期中不断生长变化的装置，其间参观者可以通过参与播种、种植与果实分享等活动来体验人与人之间的互动。

安住 · 杨梅竹斜街改造纪实与背后的思考

In 2016, the View Unlimited team's project at 66-76 Yangmeizhu Xiejie was chosen as one of the exhibits representing China at the 15th International Architecture Exhibition.

The outdoor installation "The Little Garden for the People" was crafted from discarded items found throughout Beijing's old neighborhoods, reflecting the living conditions of residents: a blend of resignation and hope. Constructed based on the layout of the alleyway at 66-76 Yangmeizhu Xiejie, the installation repurposes everyday waste collected from hutong residents. These items, once integral to daily life, carry the aspirations of those who used them. Interpreting the stories within these discarded items, we connect with them, narrating the tales of the alleyway. Within the installation, five sets of video players showcase unedited footage captured in the week leading up to the exhibition, portraying the lives of five households in the alleyway and scenes from Flora Cottage. Additionally, two sets of live-streamed videos provide real-time glimpses of the alleyway, allowing visitors and residents in Beijing to experience the exhibition firsthand, including the opening ceremony of the Chinese Pavilion.

Throughout its six-month exhibition period, this installation underwent constant growth and change. Visitors engaged in activities like planting seeds, tending to the garden, and sharing the fruits of their labor, experiencing the interactive nature of human connection.

It's a dynamic work that's always evolving, imbued with the lively rhythm of growth and transformation.

第二部分　展览传播

（一）何以为家

——关于综合装置"安住·平民花园"
（HOME·Communal Garden）的创作缘起

黄海涛

无界景观设计团队艺术总监

1.

2016 年，无界景观工作室"杨梅竹斜街 66—76 号夹道公共空间改造"项目入选第 15 届威尼斯双年展中国国家馆，这是我们 2015 年夏以志愿者身份参与的一个公益项目，是一个关于"如何让共同使用一个狭长夹道的 5 户人家打开家门，和睦交往，建立共享社区"的构想和实践。

无界景观创始人谢晓英及其团队经过多次挨门挨户、点对点的调研之后，提出为使用同一夹道的 5 户人家建立共享花草堂，通过互助种植改善邻里关系，进而实现"社区价值共享"甚至"小经济共同体"的设想。

安住·杨梅竹斜街改造纪实与背后的思考

在这个项目中，无界景观设计团队将自身定位为项目推动者，5 户人家作为设计者与团队共同制定方案。

为在威尼斯双年展上表达这个项目，我们决定在军械库中国馆室外花园按照杨梅竹斜街 66—76 号夹道的空间和比例搭建大型种植装置，使用"正负像"做法，将夹道空间从建筑群中提取出来实体化，融合现成品、种植以及多媒体等手段与展览现场观众和杨梅竹斜街 66—76 号夹道居民们共同分享我们的故事。

这是一个观众可以通过种植参与创作的共享装置，从开展到撤展共 6 个月的时间里，时刻处于变化中。

装置最初名称为"安住·HOME"，后由我们的策展顾问、中国第一届威尼斯建筑双年展策展人王明贤先生为其命名为"平民花园"。

"安住·平民花园"（HOME·Communal Garden）是团队集体创作作品，是由 20 余位年轻设计师共同完成的。此外，这个团队还额外承担了第 15 届威尼斯国际建筑双年展中国国家馆的统筹和整体搭建。

2.

由于在很"格式化"的公有制集体住宅和营区中长大，所以"老北京"对我来说曾经只意味着皇家园林，而老旧街区则一直是一个巨大的、模糊的灰色地带，对所谓"京味儿"的感受更多来自老舍的小说、北京人艺的话剧和几部电影。

大学之后才得以真切地生活在灰瓦屋顶、老的榆树、枣树和鸽子群的老城巷陌之中。

以我们学院为中心，方圆 1 千米之内的院落里住着时任的国家主席、"军队总长"、民初大军阀后裔、清朝末代国舅、国民党降将的儿孙、低保贫民、劳教释放人员……老城区居民生活的丰富与多样让人阅读不尽，像是永不落幕的即兴戏剧。那时在胡同中迷路是常态，而且似乎每次走出校门都会有新鲜的热闹可看。直到现在回到那里仍会走错方向，好像也永远走不出那个迷宫了。

我的指导老师孙家铨先生曾为我们设计过两个学期的室外课程，课题名称我已经忘记了，内容是研究、观察和记录北京的城市面貌，地点是环绕学院的几条胡同和前门大街。

关于胡同，我们先系统学习了典型四合院的形式，之后再走进现场——主要是杂院（独门独院一般是进不去的）。通过学习与观察，我们逐渐能够在这些历经翻建、改建和私搭乱建后形成的自由院落空间中有意识地去对（在时间中缓慢叠加与沉积的）历史肌理进行识别和提炼。

大杂院中的每家每户几乎都在处心积虑（但是很有分寸）地向公共空间扩张着，邻里间维持着有趣的、既相互制约又相互依存的微妙平衡，"杂院"是特别有弹性与活力的空间。

居民们在院落中堆积的生活废弃物是我们的记录重点，这些裸露在室外的器物精彩地储存并传达着每家每户内部的"私"信息，无声地讲述着每个家庭的故事。

我的毕业论文是关于杂院的，只是作为一个混日子的学生，现在已经不记得论文的题目和具体内容了。

学生时代至今，目睹了北京城史无前例的极速膨胀，见证了各种城市资源的日益紧张。大量原住民的生活仍然没有能够因城市的发展而得到相应的升级；老城新城间、新旧生活间斩截的反差加剧了生活的焦虑感，太多的不确定性导致很多原住民处于无路可退的"困"的状态中,遑论宜居了。这也是我们的项目之所以名为"安住·Home"的初衷。

城市发展引发的各种巨变让我们必须不断面对选择的焦虑。作为原住民不得不时时重新校对"家园"的坐标系,调整自己的精神"籍贯"。这不仅仅是老北京原住民需要面对的问题。

可喜的是现在的北京老城区虽破败却珍贵,俨然成为了"金矿",被各种力量以各自的方式圈占、觊觎和珍惜着。由于大量的老居民不得不维持着几十年前的生活方式,使得这一困境反而因契合了人们对"老北京"的想象而成为了"活的文化遗产",产生了很大的商机。居民们的生活场景成为可以被围观的、尴尬的人文景观,这是矛盾,然而也是机遇。

在调研中我们领略了居民们的生活技巧、生存智慧以及对美好生活的憧憬,使得大家看到了老旧街区的无奈中蕴藏着的无限潜力和可能性。

作为设计师是无力改变自己和他人的根本困境的,但是和前几个十年相比,现在的我们有了更多元的选项。在老城区改造项目中，作为设计师至少可以和居民们共同投入积极的行动中去，梳理与整合局部资源，激活居民的创造力，低成本地创造"生"的价值，力所能及地建立老城区与新生活的连接，让生活暂时地、部分地先美好起来。

感谢晓英慷慨地提供这次机会让我和北京老城再续前缘，继续孙先生留下的作业。

人真的不能两次踏入同一条河。

借郁达夫先生的一句话，就把这次作业作为我们"在京华尘土中相遇的纪念吧"！

2018.7.16

What Is Home?

Regarding the Creative Origins of the Integrated Installation "Home·Communal Garden"

Huang Haitao, 2018.7.16

1.

In 2016, View Unlimited Studio's "Public Space Renovation of 66-76 Yangmeizhu Xiejie Alley" project was selected for the 15th Venice Biennial Exhibition China Pavilion. This is a public welfare project which we participated in as a volunteering party in early 2015. It is both a concept and the practice of "how to encourage five families occupying a narrow alley to open their doors, communicate harmoniously and establish a shared community".

After many door-to door and point-to-point surveys, Xie Xiaoying, the founder of View Unlimited, and her team put forward the idea of establishing a communal garden for the five families sharing the same alley. The aim of this proposal was to improve neighborhood relations through mutual cultivation and realize the ideal of "community value sharing", or perhaps even a "small economic community".

During this project, the View Unlimited design team positioned itself as an agent of change with regards to the project. The five families took on the role of designer and worked together with the team to formulate plans.

In order to articulate this project at the Venice Biennial Exhibition, we decided to build a large-scale planting installation in the outdoor gardens of the Arsenale where the China Pavilion is located. The installation was constructed according to the dimensions and proportions of 66-76 Yangmeizhu Xiejie alley, and applied the concept of "positive and negative images" to extract space from the building complex to be materialized. By integrating ready-made products, garden installation and various multimedia, we share our story with both the audience at the exhibition site and the residents of the 66-76 Yangmeizhu Xiejie alley.

This is an interactive installation that audiences may participate in the creation of through the act of planting, and which is constantly changing over the six-month period from initial launch to dismantling.

The installation was initially named "HOME", and was subsequently renamed "Communal Garden" by our curator, Mr. Wang Mingxian, who was the curator of the first Venice Biennial Architecture Exhibition in China.

HOME· Communal Garden is a collective work which was created by more than 20 young designers. In addition, the team also undertook the overall planning and construction of the Chinese National Pavilion at the 15th Venice Biennale of Architecture.

2.

Growing up in a largely "homogenous" public-owned collective housing and barracks, for myself Old Beijing used to conjure images of the royal gardens. In contrast, the Old Beijing area has in reality always been an enormous and obscure grayish area. The feeling of the so-called "Beijing flavor" is more likely to be found in Lao She's novels, Beijing People's Art Theatre and a variety of movies.

It was only after college that we were able to live in the old city alleys, amidst the grey tile roofs, old elm trees, date trees and pigeon colonies.

Based on our campus location, the President of the State at that time, the General Commander of the army, the descendants of great warlords from the early Republic of China, descendants from the late Qing Dynasty, the descendants of Kuomintang generals, as well as poor people with low living standards and released personnel from reeducation through labor programs all lived within a 1 kilometer radius. The richness and diversity of the life of residents in the old urban areas was bountiful, akin to myriad impromptu dramas that never ended. During that time, it was common to get lost in the alleys, and it seemed that every time we went outside of school there would be a fresh new discovery waiting for us. Even now, when I visit there I still find myself going in the wrong direction, like a maze that I may never escape from.

My instructor, Mr. Sun Jiaquan, designed two semesters of outdoor courses for his students. Although the title of the subject is long forgotten, the content of the courses was to study, observe and record the urban features of Beijing, based on several Hutongs and nearby Qianmen Street which were situated around the college.

With regards to the Hutongs, we first systematically studied the form of typical courtyards, and then entered the locations - mainly courtyards shared by many households (privately held courtyards were generally not accessible). Through study and observation, we were gradually able to consciously recognize and refine the historical texture (slowly superimposed and deposited over time) present within the free courtyard space and formed after multiple instances of reconstruction and private construction.

Almost every household in the courtyard was deliberately expanding into the public space (but modestly), and the neighborhood maintained an interesting dynamic, featuring an interdependent and delicate balance. The courtyard was a particularly flexible and dynamic space.

Residents accumulated a variety of waste in the courtyard, which was one of the focuses of our records. These bare outdoor objects represented wonderful mechanisms to store and convey the "private" information of each household, silently telling the story of each household.

My graduation thesis focused on courtyard with multiple households, but seeing as I was an idle student, I can't recall the exact title or specific content of the thesis now.

From my schooldays until now, we have witnessed an unprecedented pace of rapid expansion within Beijing city and increasing constraints on various urban resources. The lives of a large number of locals have not witnessed any improvements whatsoever in spite of the development of the city; the disparity and contrast between the old and new cities has aggravated the anxiety in their lives, and this high degree of uncertainty has led to many locals existing in a "difficult" state, let alone a livable state. That's why our project is called "Home".

Various significant changes caused by urban development have led these residents to constantly face the anxiety of choice. As local residents, they must revise the geographical coordinates of their "home" from time to time and adjust their spiritual "home place". This is not just a problem faced solely by the residents of old Beijing.

Fortunately, although the old urban area of Beijing is now dilapidated it is still very valuable. It has become a "gold mine", which has been occupied, coveted and cherished by various stakeholders in their own ways. Because a large number of the old residents have no choice but to maintain a lifestyle from decades ago, this dilemma has become a "living cultural heritage", which corresponds with the people's imagination of "old Beijing", and has subsequently generated significant business opportunities. It is a contradiction, but also an opportunity in that the scenes of residents' daily lives have become an embarrassing human landscape that can be observed and watched.

When conducting the survey, we appreciated the skills applied by residents in their daily lives, their survival wisdom and an inherent longing for a better life. Indeed, this displayed the infinite potential and possibilities hidden amidst the helplessness of the old city alleys.

As designers, we cannot change the fundamental dilemma faced by ourselves and others, but now we have more options available to us than in the previous decades. As a design team, we can at least join the residents in making a positive action in the old city urban renewal project by combing and integrating local resources, inspiring the creativity of the residents, creating a high quality of life at low cost, and establishing a symbiotic link between the old urban area and modern life to the extent that we may. In doing so, we may improve the quality life in pockets of the old city in phases.

Thank you Xiaoying for generously offering this opportunity for myself and the old city of Beijing to continue the work passed on by Mr. Sun.

Those that genuinely persevere shall not cross the same river twice.

In the words of Mr. Yu Dafu, we shall make this project a "commemoration of our meeting in the dust of Beijing".

2018.7.16

安住·杨梅竹斜街改造纪实与背后的思考

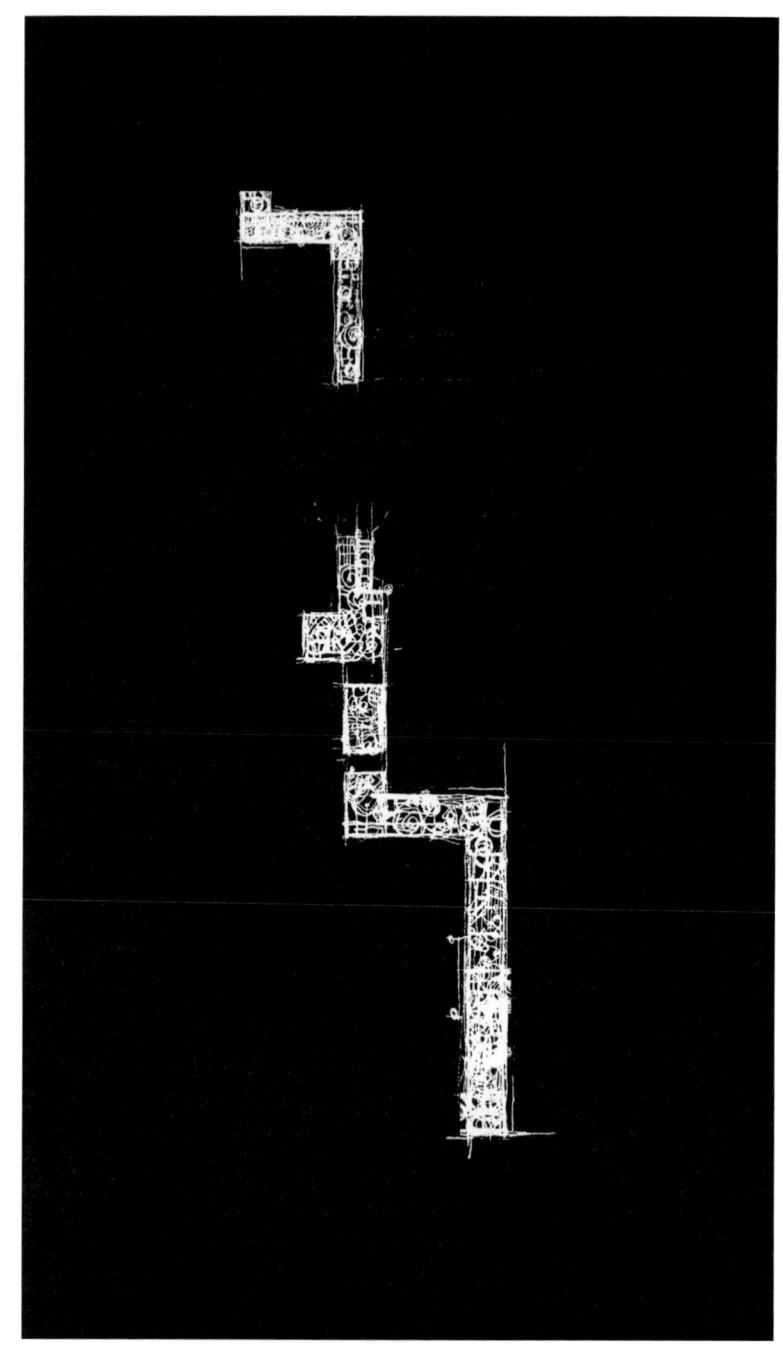

根据 66—76 号夹道实景生成的设计草图
Design sketch based on the actual generated scene of the alley of Nos. 66-76

第二部分　展览传播

夹道实景
The actual generated scene of the alley of Nos. 66-76

安住·杨梅竹斜街改造纪实与背后的思考

在军械库中国馆室外花园，按照杨梅竹斜街 66—76 号夹道的平面，运用当代艺术的手法，严格按比例搭建了大型综合装置。

In the Armory of the China Pavilion outdoor garden, a large-scale comprehensive installation has been constructed based on rigid proportions by means of contemporary art and in accordance with the plan of nos. 66-76 Yangmeizhu Xiejie alley.

第二部分　展览传播

具体形式上借鉴摄影中的"正负像"作法，将围合挤压出夹道空间的房屋建筑虚化，将夹道空间实体化，融合现成品、多媒体与种植的手段来和观众分享我们的实践心得。

Referring to the method of "positive and negative images" in photography, the houses and buildings are blurred, provide a materialization of space of the actual alley, and integrate the means of ready-made products, multimedia and planting to share our practical experience with the audience.

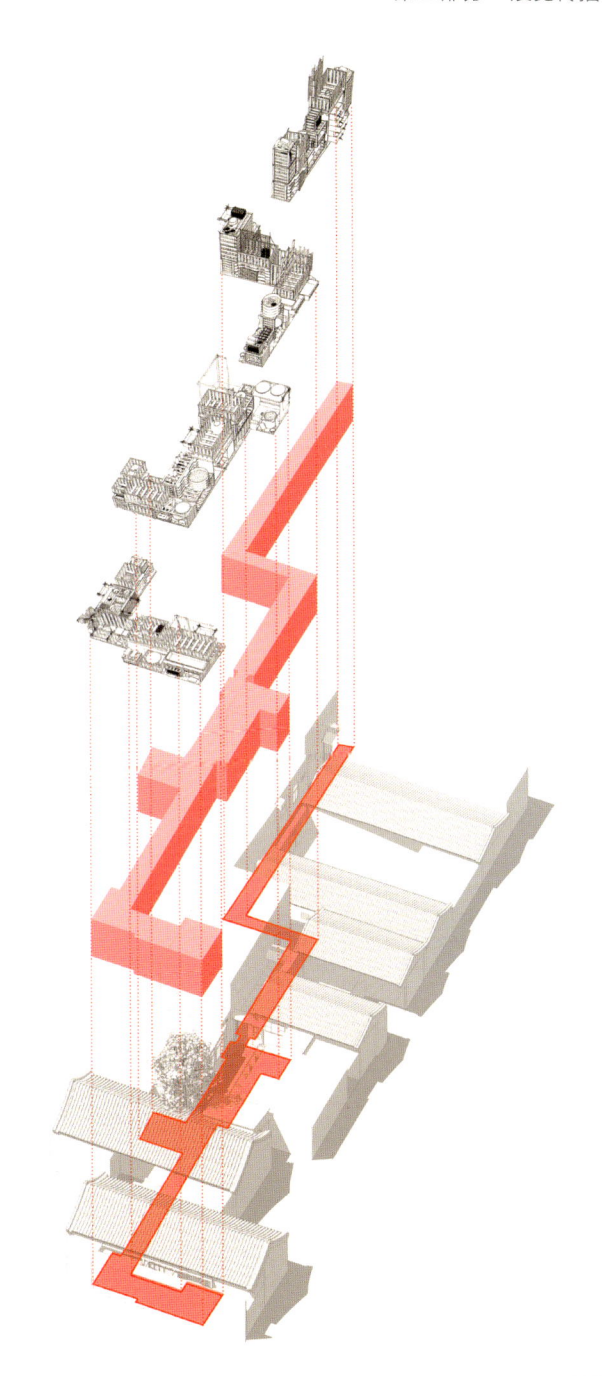

安住·杨梅竹斜街改造纪实与背后的思考

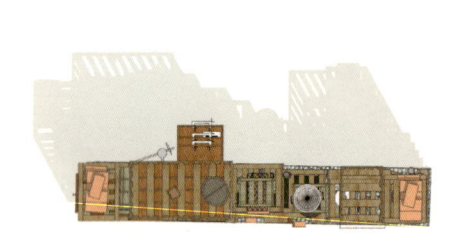

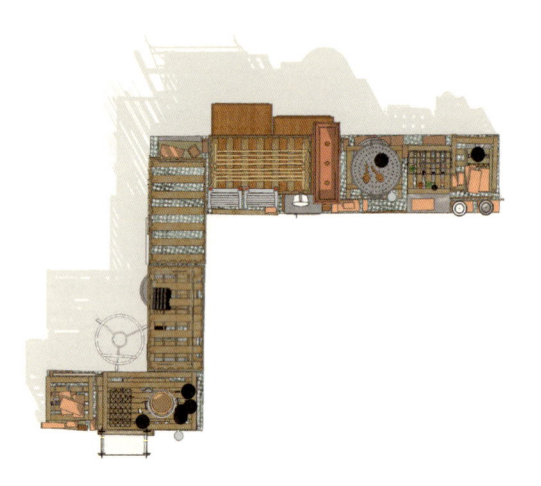

布展效果图
Render of the exhibition

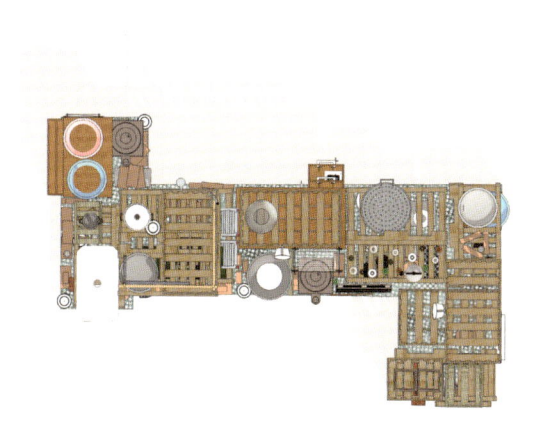

第二部分　展览传播

安住·杨梅竹斜街改造纪实与背后的思考

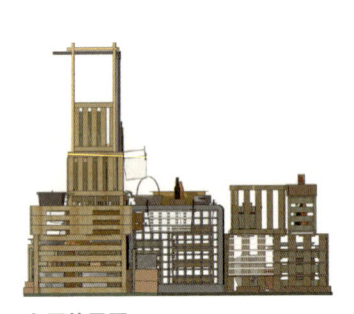

布展效果图
Render of the exhibition

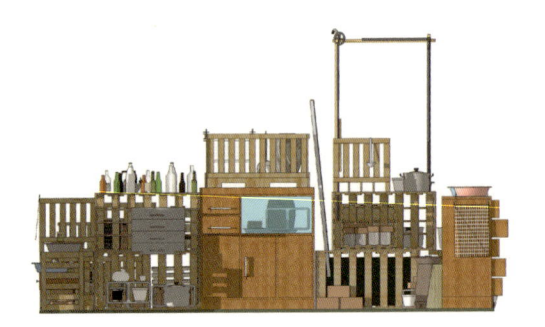

第二部分　展览传播

安住·杨梅竹斜街改造纪实与背后的思考

布展效果图
Render of the exhibition

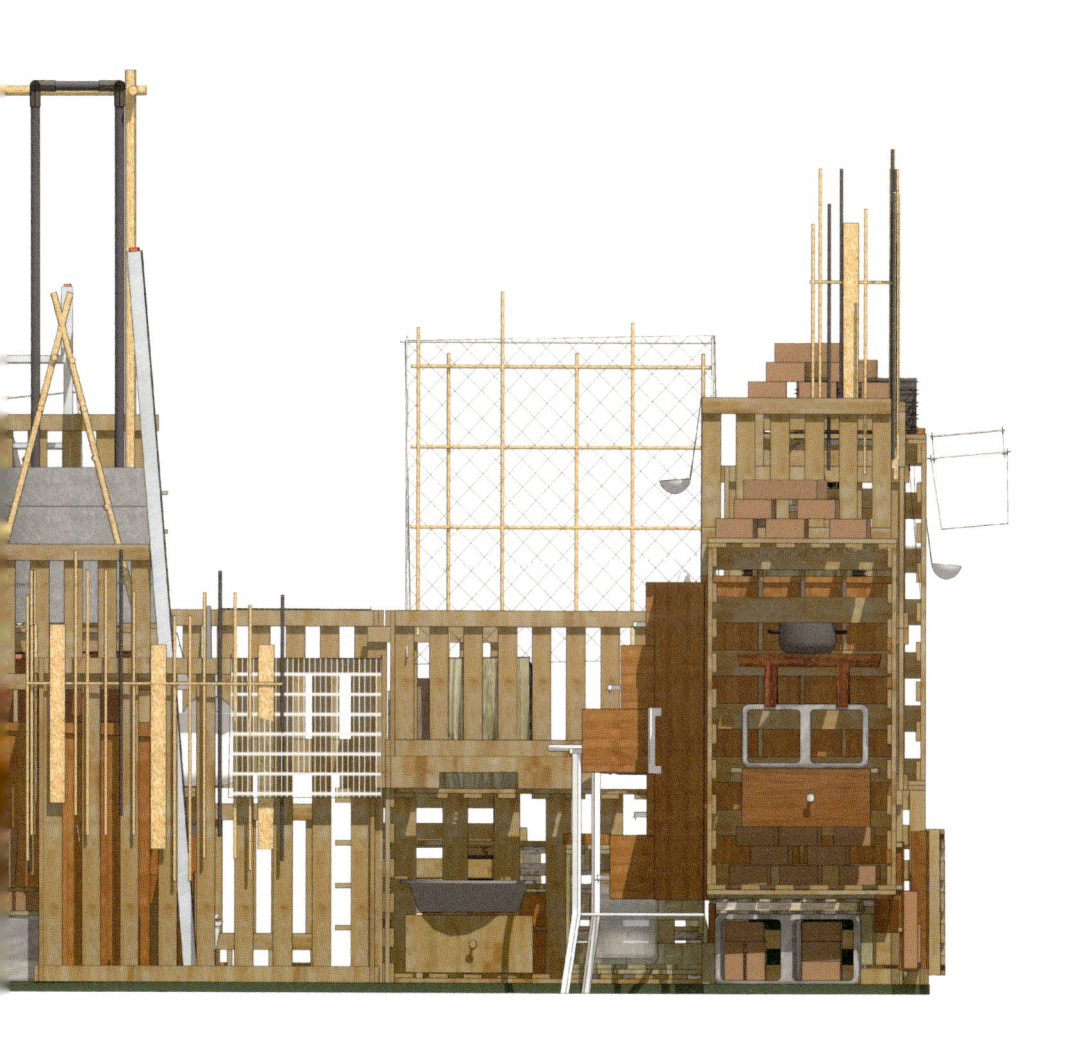

第二部分　展览传播

安住 · 杨梅竹斜街改造纪实与背后的思考

（二）作品简介

预搭建
Pre construction

北京
Beijing
2016.2

第二部分　展览传播

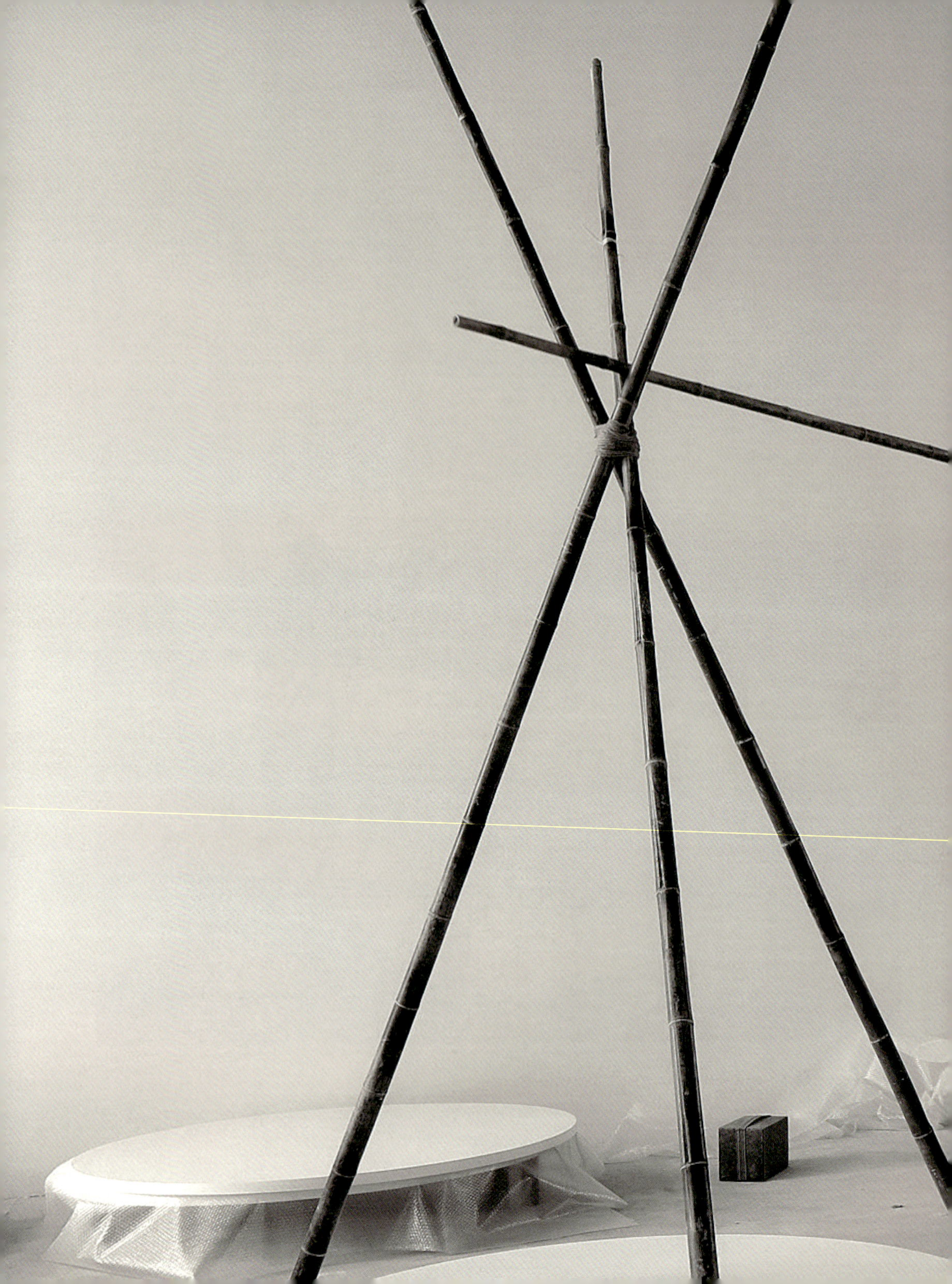

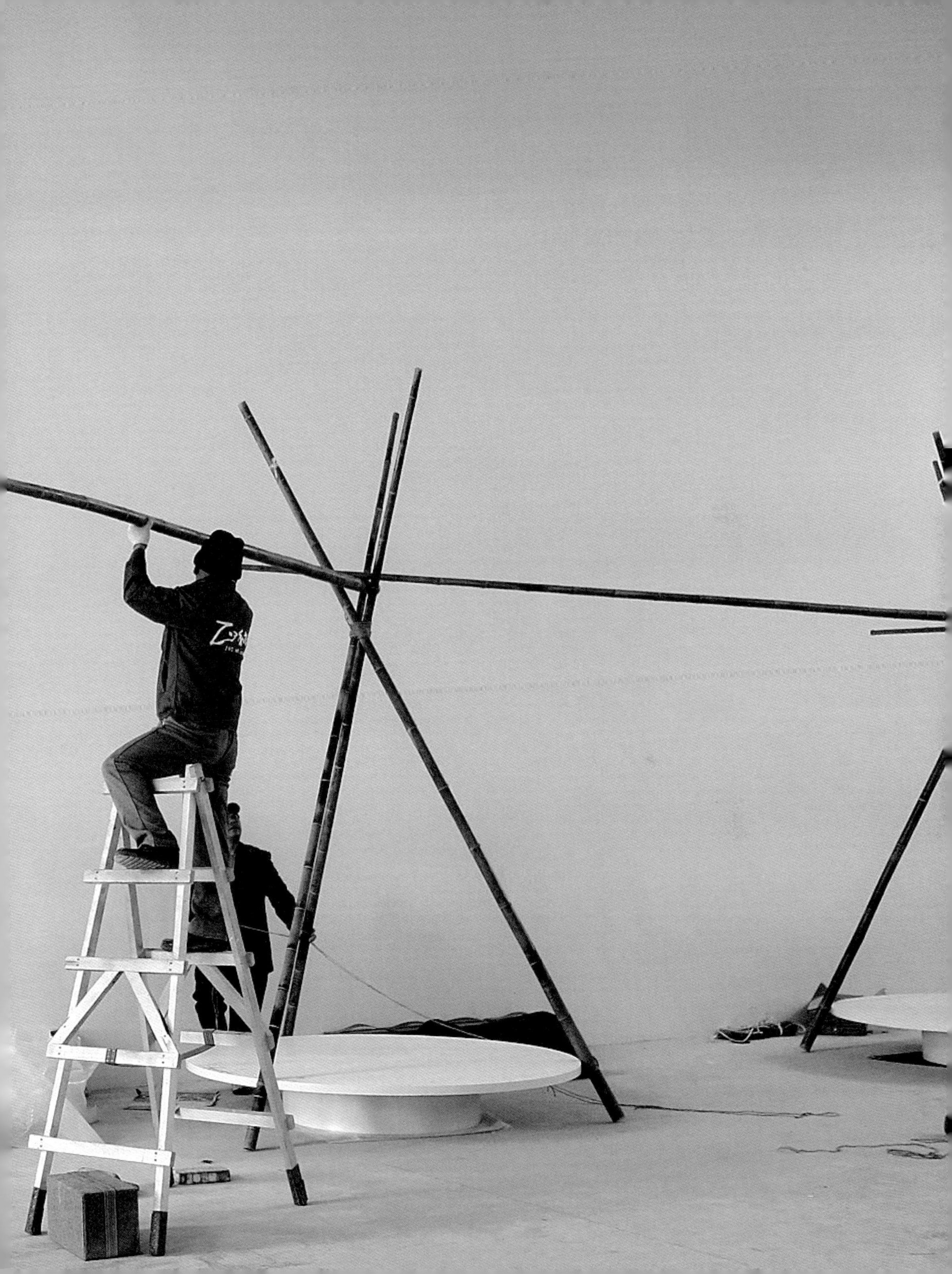

安住 · 杨梅竹斜街改造纪实与背后的思考

预搭建现场
Site of pre construction

第二部分　展览传播

安住·杨梅竹斜街改造纪实与背后的思考

第二部分　展览传播

安住·杨梅竹斜街改造纪实与背后的思考

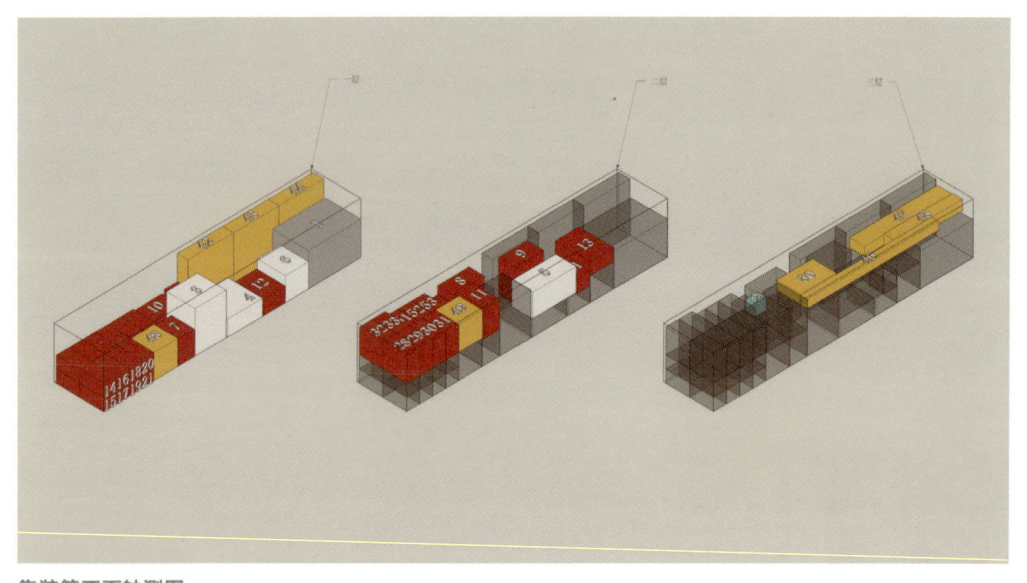

集装箱正面轴测图
Front Axonometrical Drawing of the Container

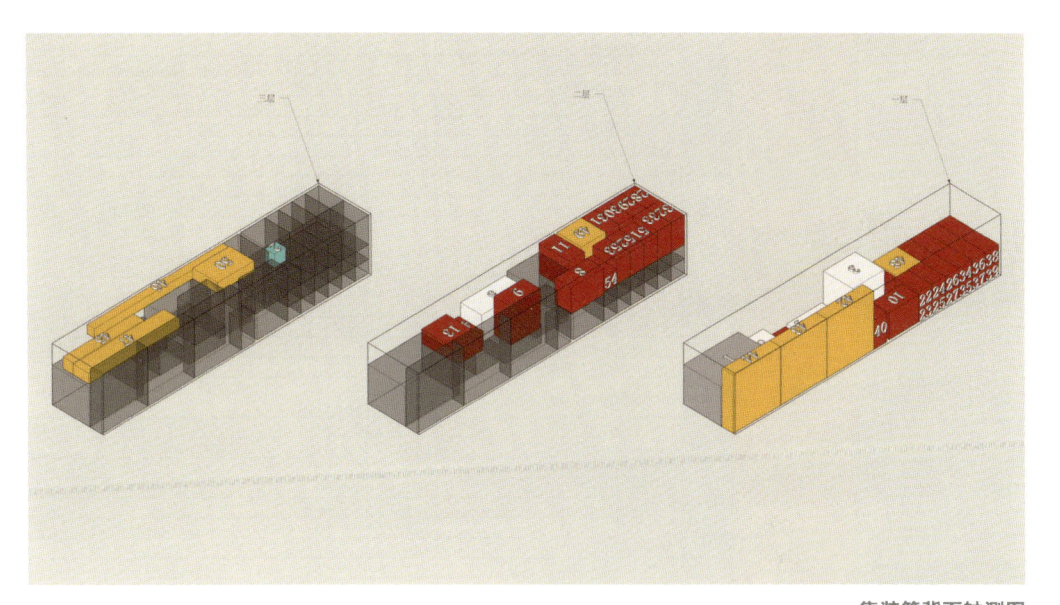

第二部分　展览传播

集装箱背面轴测图
Axonometrical drawing of back of container

155

安住 · 杨梅竹斜街改造纪实与背后的思考

名称	箱号	尺寸	物品号
宋群	1	2440*1670*1480	见宋群清单
众建筑	2	510*410*500	见众建筑清单
润建筑	3	1310*1310*1800	见润建筑清单
	4	1600*1600*640	
	5	950*950*1900（1020）	
	6	1400*1150*1050	
无界景观	7	1100*800*1050	144，160（18），158（7），161（17），168（2），169
	8	1470*720*990	149，165（39），
	9	1320*770*1440	147，165（？），
	10	1370*800*1010	146，165（？），
	11	1150*710*1350（1150）	148，165（？），
	12	1000*650*1340（650）	145，165（？），
	13	1600*1100*550	97,98,117，装不下锯断了），155,163,164,236（3），
	14	1100*600*550	4，7（2），18，22（1），27，34，36（1），66,67，71
	15	1100*600*550	6，8,16,43,44,46，48，49（2），56,68（7），71（2）
	16	1100*600*550	7（1），9（1），12,22（1），23,25,26,29,30,31,35,36
	17	1100*600*550	32，59,63,64,68（2），170（12），171（46），172（25
	18	1100*600*550	3，10,11,13，33,39，47，61，78，120,121,122,123,12
	19	1100*600*550	9（1），33，60,62,124,168（7），174（8），185,189
	20	1100*600*550	99,173,174（1），175,176,184,158（10），165（1）
	21	1100*600*550	40，70（1），77,100,104,196,161（10），165（1），
	22	1100*600*550	160（2）161（46），213（1），214（1），
	23	1100*600*550	5，14,20,28,37,70（1），74，80，81,82,101,106，1
	24	1100*600*550	75,158（7），167（1），178,230，165（3），213（1
	25	1100*600*550	160（2），161（40），213（1），214（1），
	26	1100*600*550	160（2），161（40），213（1），214（1），
	27	1100*600*550	160（2），161（41），213（1），214（1），
	28	1100*600*1100	165（20），224（2）
	29	1100*600*1100	108，140（2），153，160（8），室内清单22（10），室
	30	1100*600*1100	21（2），42，103，107，115（1），135,136，152，小
	31	1100*600*1100	76，111,112,113，116，140（2），226（2），228，231
	32	1100*600*1100	1，24，41，65，69,83,127，128,129,130,131,132,133 （1），165(12)，红水壶
	33	1100*600*1100	2，16,17，21（1），45，79,114,115（2），118,119,15
	34	1100*600*550	165（7），213（1），室内清单28（1）
	35	1100*600*550	234，235，165（1），213（1），室内清单28（1），
	36	1100*600*550	151，165（3），213（1），室内清单28（1），
	37	1100*600*550	227（1），165（7），213（1），214（1），
	38	1100*600*550	159（5），165（10），209,210,211，213（1），
	39	1100*600*550	226（1），227（2），165（3），213（1），
	40	1100*600*550（1100）	室内清单28（1），165（8），213（1），
	41	1100*600*550	52，109（拆了），110（拆了），166，237，
场馆室内	42	2100*2200*500（2200）	见室内清单
	43	2100*2200*500（2200）	见室内清单
	44	2100*2200*500（2200）	见室内清单
	45	3200*460*440	
	46	6100*320*350	见室内清单
	47	4100*400*700（400）	154，162，见室内清单
	48	1100*1100*1100	38，165（19），224（1），室内清单28（16），
	49	1100*1100*1100	50，51，54，55，57，58,84,85,86,87,88,89,90,91,92,93，
	50	1650*400*1100（400）	165（3），215（5），216（20），217（20），218（1
新增加	51	1100*600*900	165（6），224（2）
	52	1100*600*900	165（4），224（2）
	53	1100*600*900	165（4），224（2）
	54	1100*600*560	165（8），224（1）

海运物品清单
List of marine goods.

变动：102（可能没了），115（改为4个，比原来多1个），21个，比原来少1个），214（改为15个，比原来少1个）个），

增加：席子(2)，陶花盆，瓷花盆，陶花盆，青花小盆，

第二部分　展览传播

），171（4），172（6），177，

席子(2)
），165（2），213（1），214（1），
，158（7），213（1），214（1），陶花盆
（1），73,142,172（19），180,181,195,158（2），165（3），213（1），214（1），
，178（2），191,192,193,197,220,221,223,225,165（1），213（1），214（1），
），213（1），214（1），瓷花盆
，161（10），165（1），213（1），214（1），陶花盆
214（1），
14（1），
187,194,165（2），213（1），214（1），青花小盆
，

丁），室内清单27,室内清单30，

电箱），232,室内清单22（10），室内清单24（变压器）（1），
，141,156,157,179,182,132,188,190,198,199,200,201,202,203,204,205,206,207,208,229,236
165（5），穴盘

9（5），219,室内清单28（20），火钳子（2）
25,室内清单26,亚克力板（2）

为4个，比原来少2个），160（改为34,比原来多9个），161（改为194,比原来多14个），213（改为
5个，比原来少15个），219（管子100米，比原来增加50米），室内清单24（改为1个，比原来少7

叶盒，红水壶，穴盘，火钳子（2），亚克力板（2）

157

安住·杨梅竹斜街改造纪实与背后的思考

展区搭建
Exhibition area to build

威尼斯军械库
The Venice Armory

2016.5

第二部分　展览传播

安住・杨梅竹斜街改造纪实与背后的思考

第二部分　展览传播

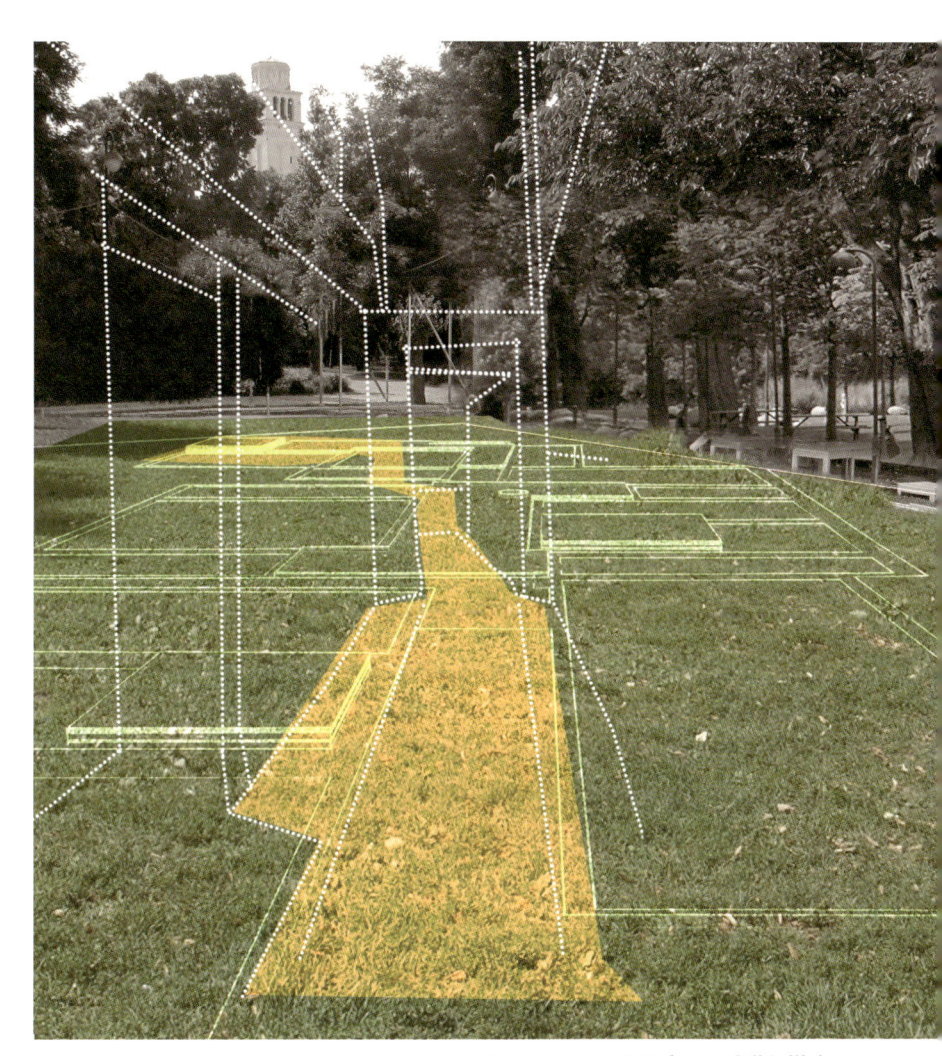

根据 66—76 号的实际尺寸进行搭建
Build according to the actual size of no. 66-76

放样
Lofting.

第二部分　展览传播

安住 · 杨梅竹斜街改造纪实与背后的思考

布展过程中
Arranging the exhibition

第二部分　展览传播

安住·杨梅竹斜街改造纪实与背后的思考

布展过程中
Arranging the exhibition

第二部分　展览传播

安住 · 杨梅竹斜街改造纪实与背后的思考

布展过程中
Arranging the exhibition

第二部分　展览传播

安住・平民花园装置作品
Home Communal Garden Installation

第二部分　展览传播

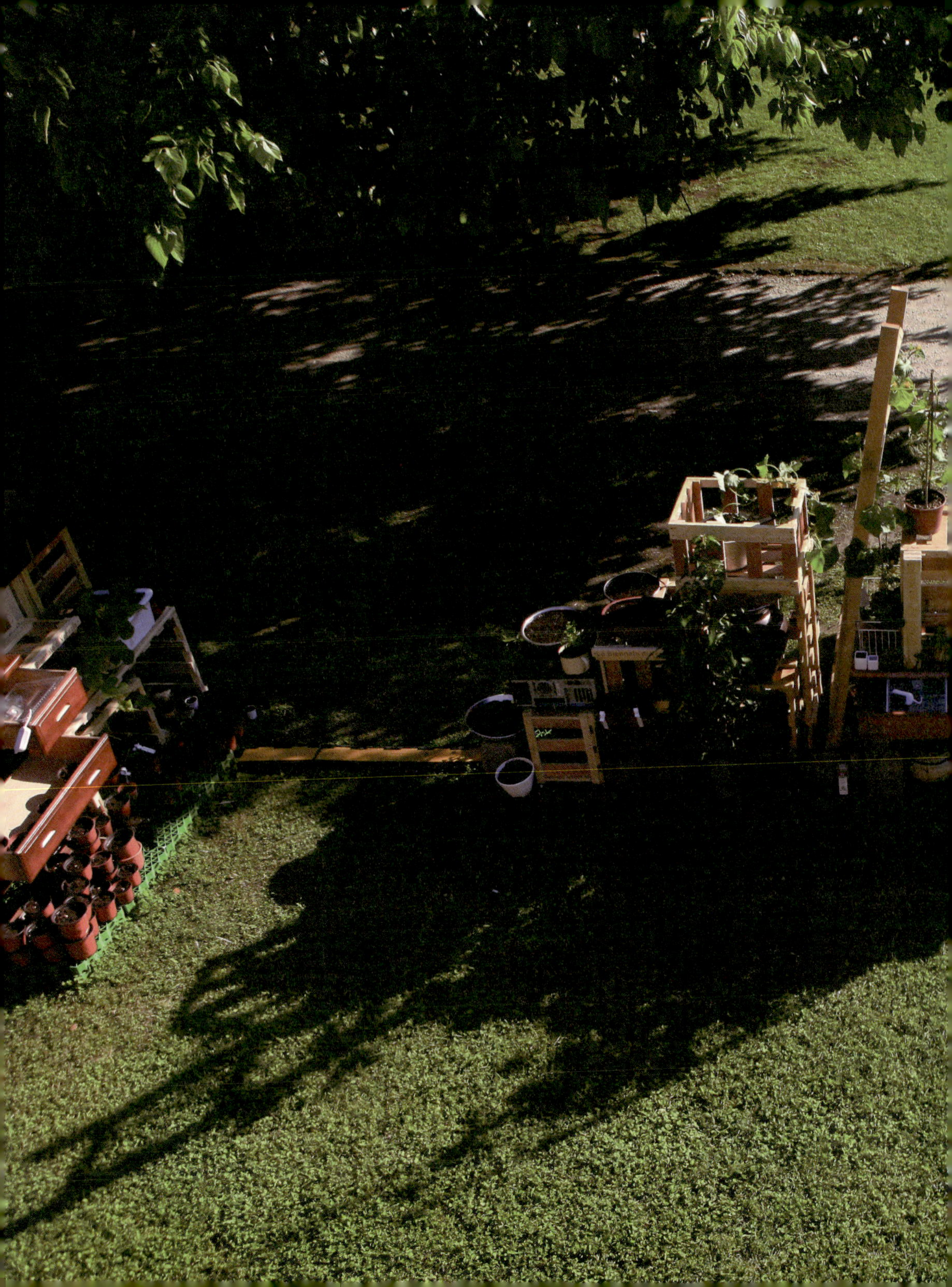

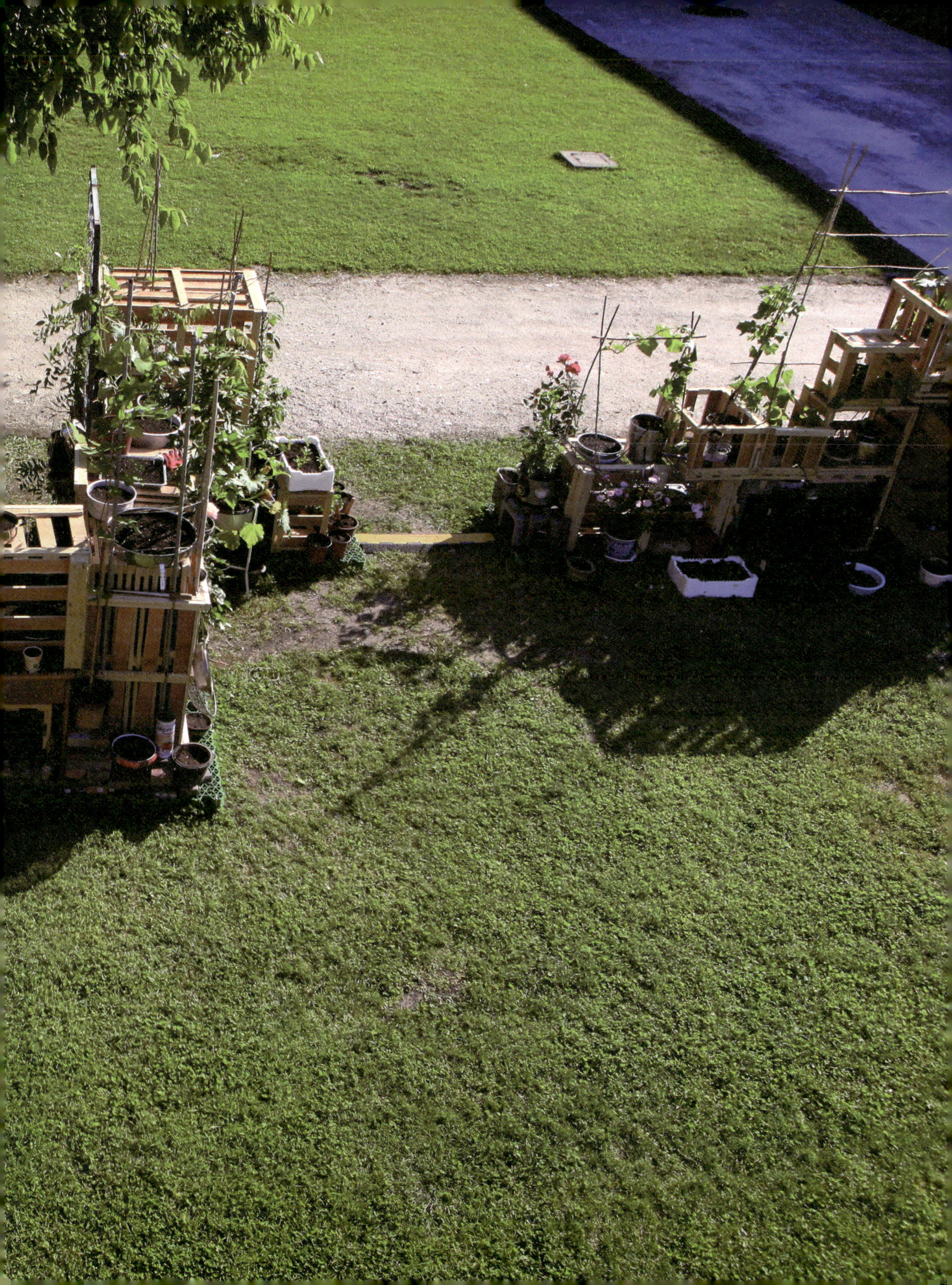

安住 · 杨梅竹斜街改造纪实与背后的思考

种植器皿来自老北京最常见的容器

These planting containers are derived from the most common containers found in old Beijing

第二部分　展览传播

安住·杨梅竹斜街改造纪实与背后的思考

第二部分　展览传播

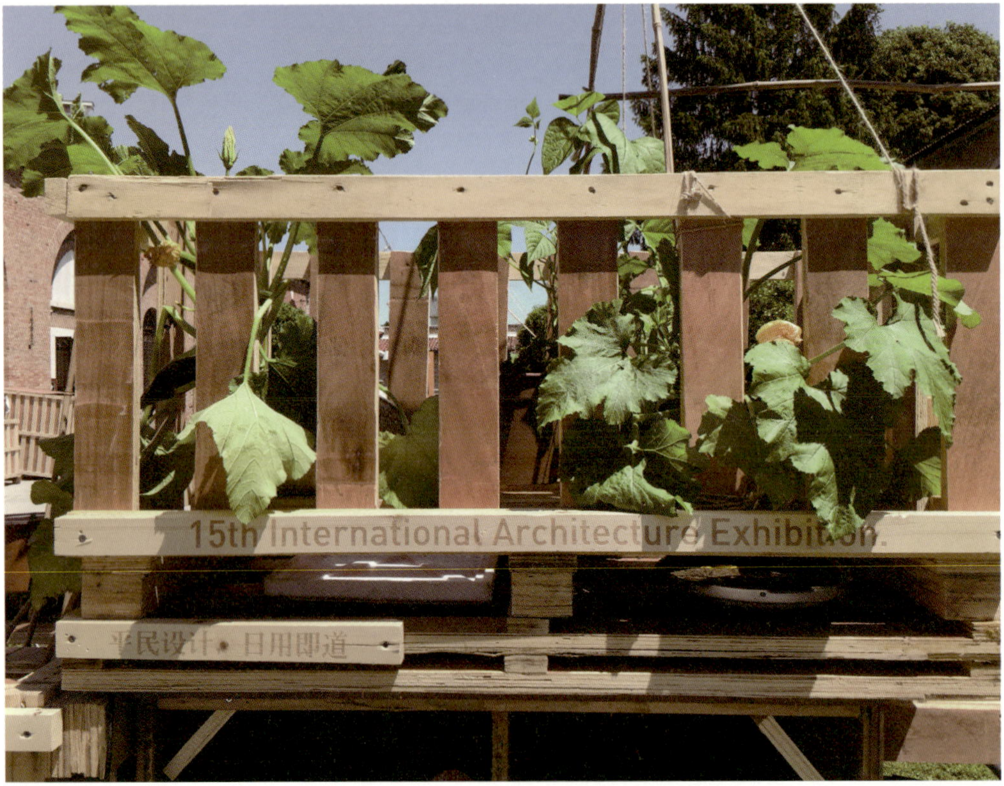
15th International Architecture Exhibition.
平民设计 · 日用即道

第二部分　展览传播

开幕
Opening ceremony

威尼斯
Venice

2016.5

PAVILION OF CHINA
PLANTING ACTIVITY
HOME · COMMUNAL GARDEN
25 MAY 2016 - 28 MAY 2016

15. Mostra
Internazionale
di Architettura
Partecipazioni Nazionali

HOME
Communal Garden
China Pavilion
2016.5.26-11.26

INVIT@tion

179

开幕式上中国文化部官员及意大利议员参与播种仪式（2016.5.24）
Chinese officials from the Ministry of Culture and Italian parliamentarians participate in the ceremony

第二部分　展览传播

开幕式当天前来体验种植的参观者
Visitors take part in planting activities on the day of the opening ceremony

第二部分　展览传播

开幕式当天前来体验种植的参观者
Visitors take part in planting activities on the day of the opening ceremony

第二部分　展览传播

安住·杨梅竹斜街改造纪实与背后的思考

直播及
夹道居民互动

Live broadcast
featuring interaction
with local residents

威尼斯 & 北京
Venice & Beijing
2016.5

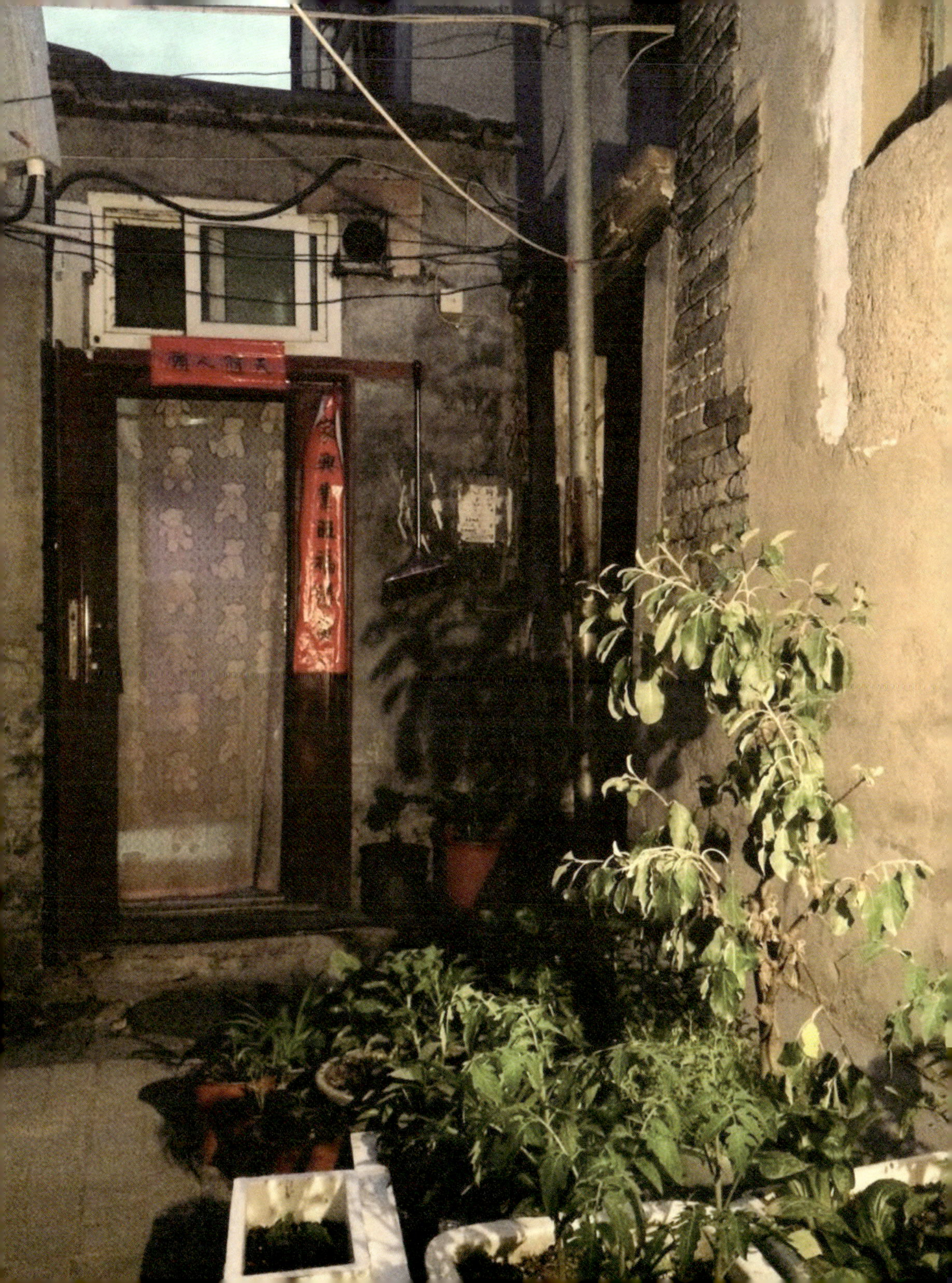

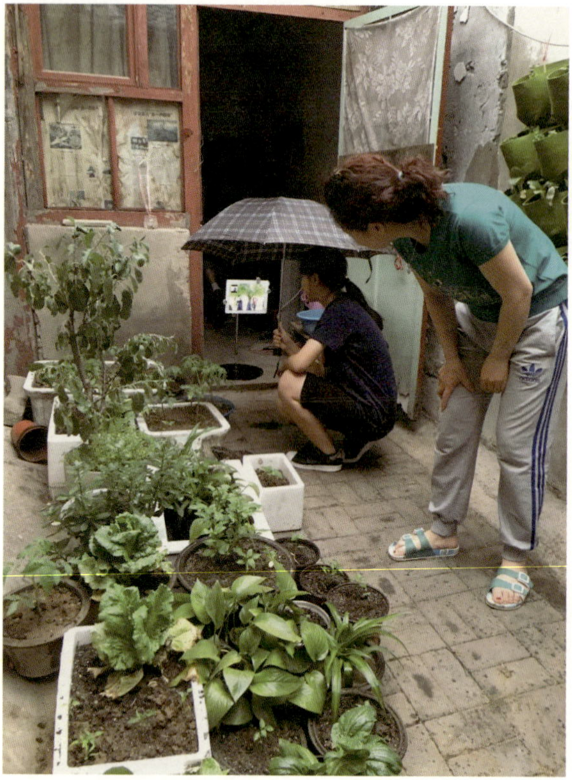

第二部分　展览传播

同一时间，不同空间，威尼斯双年展参观游客与北京"胡同花草堂"居民通过网络连线进行互动
At the same time, despite the significant distance, visitors to the Venice Biennale were able to interact with residents of Beijing's "Hutong Floral Cottage" through a network connection

安住·杨梅竹斜街改造纪实与背后的思考

第二部分　展览传播

同一时间，不同空间，威尼斯双年展参观游客与北京"胡同花草堂"居民通过网络连线进行互动
At the same time, despite the significant distance, visitors to the Venice Biennale were able to interact with residents of Beijing's "Hutong Floral Cottage" through a network connection

191

安住·杨梅竹斜街改造纪实与背后的思考

第二部分　展览传播

（三）安住·平民花园

——第 15 届威尼斯国际建筑双年展中国馆参展项目

谢晓英

中国城市建设研究院
无界景观工作室主持设计师

第 15 届威尼斯国际建筑双年展于 2016 年 5 月 28 日正式开幕。本届双年展的策展人是 2016 年普利策奖获得者，智利建筑师亚力杭德罗·阿拉维纳（Alejandro Aravena）。他为本届建筑双年展策划的主题是"来自前线的报告"。这一富有激情的主题自然引来各国建筑师与策展人的响应，并以各自的方式对这一主题进行阐述。英国国家馆策展人雅克·塞尔夫（Jack Self）认为这一主题极具挑战性，阿拉维纳不仅消解了建筑学的普遍性前提，也提出了作为社会人的建筑师的人道主义角色问题，是一次建筑学的范式转换。

1. 主题阐释

"前线",这种看似欧洲左翼的表述方式在 1990 年之后已经不多见了,尤其是在建筑领域内。这一激进的标语式的主题之所以引起广泛的反响,既与阿拉维纳的个人经历、背景和观点主张相关,也与冷战结束后的全球化扩张中发展中国家的现实境况有关。在包括中国在内的发展中国家里,经济繁荣的背后是社会阶层分化与贫富差距的扩大,而这一现象最直观的体现就是建筑与人居环境。

阿拉维纳在此次策展陈述中对新"自由主义""个人主义意识形态的批判",以及"平等""集体"等那些我们曾经熟悉的概念的张扬都反映了作为建筑师的策展人所思考与关注的焦点并不在建筑本身,而是如何以建筑为工具去解决非建筑的问题,也即在建筑设计、城市设计、景观设计等之外的社会问题。正如他不久前在上海接受访谈中所说的:"当建筑师开始关心社会的时候,他们就开始不在乎做一些'差建筑',甚至不把自己当成建筑师。用建筑作为工具,解决建筑以外的事情,因为建筑在建筑之外会更有力量。"

阿拉维纳在其策展陈述中表示：希望不同国家的建筑师、策展人以各自的方式思考和定义"前线"的概念，并通过实际案例展示各自不同情境下所取得的哪怕是"一毫米"的改进。

在此次参展的 54 个国家馆中，英国、新加坡、德国、芬兰、斯洛文尼亚等国家馆的展览主题都与"家"（Home）相关。他们分别从不同的视角阐释了"家"与人、与社会、与身份认同、与建筑、与时间、与难民等问题的关系，并通过不同的案例提出了建设性的理念与方案。

本届威尼斯建筑双年展中国国家馆的展览主题是："平民设计，日用即道"。展览以传统手工艺、城市改造和乡村建设三个部分 9 个案例展示了多位建筑师、景观建筑师在各自的设计实践中对中国当代建筑、环境以及社会问题的思考与解决方案。其中由"无界景观"参展的一个实际案例"安住·平民花园"（HOME·Communal Garden）同样以"家"为核心探索了当代中国特殊情境中的人与人的关系问题以及社区重建的可能性。

2. 花草堂

"安住·平民花园"由一组日常器物与植物、视频装置组成，与建筑师朱竞翔先生设计的主题为"斗室"的拼装式建筑在中国国家馆外面的花园草坪上并行展出。该案例是"无界景观"设计团队于 2015 年参与并仍在进行中的一个公益项目。

该公益项目是设计团队自杨梅竹斜街改造项目完成后的一次延伸设计。为解决杨梅竹斜街改造前缺乏乔木类植物、绿化环境差的问题,我们在环境改造设计中加建了街道与居民房屋衔接的花池用来种植观赏花及灌木,以增加街道的绿化量。但是所有这些出于美化环境目的的举措并没有受到多数居民的支持。这一现象引发了我们对公共空间、公共秩序等一系列问题的再思考。我们没有将上述这些现象的成因简单地归结为国民素质的低下或公共道德的缺失,而是通过实地调查与分析,从社会学的视角对这些现象背后的特定社会情境与特定人群之间的社会关系进行研究。研究的结果不仅使我们对那些社会底层民众的生存境遇有所体悟,也改变了我们对设计的一贯理念,即设计应该去适应设计的对象而不是改变;设计不应以普遍的形式强加于那些哪怕看似丑陋的特殊对象,这种对异质性的恐惧与排斥恰恰是同质化的根源。正是基于这种观念,我们于 2015 年开始了"花草堂"建设的实验。

这个被命名为"花草堂"(Flora Cottage)的平民花园位于北京市大栅栏街区杨梅竹斜街 66—76 号的一条长 66 米、最窄处只有 1 米的夹道中,由五户常住和暂住的人家以及外来务工人员共同使用。这是一个非常典型的以"大杂院"为主的北京城老旧街区,一个由各种历史沿袭与近期城市改造、文化街建设等因素构成的复杂的人居环境。

在一个破败的夹道中建设一个共同种植空间，其主要目的并不是为这五户居民建立一个休闲、娱乐、赏花的场地，而是通过种植活动为这五户有着不同背景、来历、彼此相邻但却少有往来的居民们建立一个交流与共享的平台。在项目的前期调查中，设计师们发现，尽管这五户居民的生活环境非常恶劣，现实中也存在着各种矛盾与利益冲突，但唯有种花、种菜是他们共同的爱好，也即种植是这些居民之间的最大公约数，是他们之间可借以相互交往的中介。

种植与绿化本是景观建筑设计的核心，它与人的视觉感受、环境、生态等都有着密切的关联。而在这一公益项目的调查中，设计师们却发现了种植的另一种社会功能，即如果加以引导，种植这种自发的个体行为可以转化为自发的公共行为；只要有一个共享的空间，就存在着形成自发社区的可能性。在这一过程中，设计师的角色也从"为平民设计"的慈善者转向"平民自建"的引导者和参与者。这种角色的转换喻示了一种设计者与服务对象之间的平等关系，也是对策展人阿拉维纳在其策展陈述中所言"建筑应作为通向平等的捷径"的回应。

对于一个以社会学意义为目标的项目展示，其展品本身已经显得无足轻重，而更重要的是让参观者知道项目的原属地正在发生什么。这组装置中用于种植的器皿——锅、碗、瓢、盆、泡沫包装盒、废弃的建筑材料等都是设计师们在北京老旧城区中收集的,是最常见的城市底层民众用于种植的"免费"器皿——那些在"环保"意义上可循环利用的废弃物,而摆放这些器物与植物的"展台"是用于包装和运输的木箱。

在这组装置中我们设置了五组视频播放器,视频内容是设计师们开展前一周录制的,没有经过任何剪辑的夹道中五户居民的生活片段以及"花草堂"实景。同时,展览现场还有两组视频是通过网络实施转播的。参观者可以通过视频实时看到夹道中的景象,而远在北京的居民们也可以通过视频看到展览现场的实况(包括中国馆的开幕式)。这样做的目的是想使这个展示更像一个报告,且是一个正在实施过程中的报告,而不是一件完成的设计作品。展示不是以艺术的方式如剪辑、特效等纪录片式的再现手法去讲述一个故事,而只是想让参观者了解这里正在发生什么。

被称作"处女花园"的威尼斯双年展中国馆的展览现场绿草如茵,环境优美,加之展品中所种植的植物、花卉生长茂盛,这个项目给人的第一印象可能就是一个花园设计。但来自"花草堂"现场的视频直播所展现的是:在一个面积不足 10 平方米的夹道空间中,种植在一些由居民捡拾来的泡沫包装盒、油漆桶等废弃物中的各种蔬菜或瓜果和少许的观赏植物。由此形成的强烈对比也使观众更容易去理解这个项目在当下中国特定情境下的特殊意义，人们才可能会意识到：展示在眼前的这堆器物与植物真的不是一个人们想象中的花园，也不是什么设计作品，更不是一件艺术品，而是一个来自现场的报告。

3. 日用即道

日用即道。日用乃日常之习性，乃日常生活之外化；"道"乃真理，乃生活意义之所在，乃终极目的。这既是中国传统哲学的一个组成部分，也是普通百姓对生活的信念。在参与由梁井宇先生主持的"大栅栏杨梅竹斜街改造项目"过程中，我们始终坚持着这一理念。但是，现实逻辑也使整个改造计划充满了矛盾与悖论。这是一个由政府主导，由投资公司运作的商业项目，包括居民搬迁、违章建筑拆除、街道铺装与建筑立面改造等。这里就涉及了资本运营与操作的问题。按照资本的逻辑：效益即道，增值即道。那么，生活于这条街道中的居民们的日常生活如何产生效益？何以增值？答案似乎只有两个选项：搬迁，或改变自身的日常生活的方式，以适应资本增值的需要，即日常生活不再是自身的目的，而是作为实现他者目的的手段；日常生活仅仅是一种被重新编码的"胡同生活"的展示。日常不存，日用何为？道将焉附？正像一位街道居民所感叹的："这个街道现在变成文化街了，和我的生活没有什么关系了。"

尽管我们深知当今世界上资本无往而不胜的现实，但是，在整个街道改造过程中我们仍然坚守着那一初衷——为平民设计，使改造工程能够最大限度地适合当地居民们的日常生活。如果把这里比作"前线"，那么，我们所做的就是以日常生活之道 VS 资本所营造的奇观社会。

"前线"在当下中国的特殊情境中意味着什么？也许正如阿拉维纳在策展陈述中所说的："在这星球上，越来越多的人们想要寻找一个适合居住的家园，然而这个过程正变得越来越困难。任何想要打破既有规则的尝试毫无例外都要遭受现实的巨大阻碍，任何尝试解决相关问题的努力都需克服日益复杂的社会变化。"

作为一个公益项目，夹道"花草堂"本应在 2015 年 10 月就可以完成，最初的设计面积也比现在这个不足 10 平方米的小小空间大很多，其中所遇到的来自各方面的阻力可想而知。作为一个社会性实验项目，许多外在的不确定因素是设计师们难以把握的，其结果是否能够达到预想的目标还需要长期的观察与不断地调整才能显现，但无论结果如何，这都是一次在设计之外的，具有建设性的尝试。

2016.6

HOME·Communal Garden

——15th Venice Architecture Biennial China Pavilion Exhibition Project

Xie Xiaoying

The 15th Venice Architecture Biennale was officially opened on May 28, 2016. The curator of this biennial exhibition is the 2016 Pulitzer Prize winner, Chilean architect Alejandro Aravena. His theme for this Architecture Biennial is "Reporting from the Front". This passionate theme naturally attracted the response of architects and curators from all over the world who have elaborated on the theme in their own ways. Jack Self, curator of the British National Pavilion, considers the theme to be very stimulating in its challenge. Alejandro not only dispelled the premise of the universality of architecture, but also raised the issue of the humanitarian role borne by architects as social people, which is a paradigm shift within the field of architecture.

Reporting from the Front

"Frontline", a seemingly European left-wing expression, has been rarely seen since 1990. This is especially true in the field of architecture. As a result, this radical, slogan-like theme has caused widespread reverberations. It is related not only to Aravena's personal experience, background and opinions, but also to the realities of developing countries in the Post-Cold War expansion of globalization. In developing countries, including China, behind the nation's economic prosperity is a widening gap between the rich and the poor, and the most intuitive manifestation of this phenomenon is found in construction and living environments.

Aravena's critique of neoliberalism and individualism in this curatorial presentation, and the promotion of the concepts of equality, collectivity and so on, which we used to be familiar with, are all a reflection of the curator's present concern and overall philosophy. The curator's focus is not on the architecture itself, but on how to use architecture as a tool to solve non-architectural problems. Namely, social problems that go beyond architectural design, urban design, landscape design, etc. As he said in an interview in Shanghai recently, "When architects begin to care about society, they don't care about doing some 'sub-standard buildings'. In fact, they may not even think of themselves as architects. Buildings should be used as tools to solve issues that go beyond architecture, because those elements are even more powerful that architecture itself."

In his curatorial statement, Aravena expressed his hope for the architects and curators from different countries to consider and define the concept of "frontline" in their own ways, and to be able to showcase improvements of even "one millimeter" in disparate situations through practical cases.

Among the 54 national pavilions participating in the exhibition, the exhibition themes of the British, Singaporean, German, Finnish, Slovenian and other various national pavilions are all related to the concept of "home". They explain the relationship between "home" and people, 'home' and society, "home" and identity, "home" and architecture, 'home' and time, as well as 'home' and refugees through the prism of different perspectives. Moreover, through a variety of individual cases, constructive ideas and concepts were proposed.

The theme of the Pavilion of China in this Venice Architecture Biennial is "Daily design, Daily Tao". The exhibition takes 9 cases which fall into the three categories of traditional handicrafts, urban transformation and rural construction, in order to show the concepts and solutions of many architects and landscape architects with regards to contemporary Chinese architecture, environment and social problems as relating to their respective design practices. Among them, a real-life case exhibited by View Unlimited titled "HOME· Communal Garden", which similarly applied "home" as the core concept, explored the relationship between people and the possibility of community reconstruction amidst the unique situation of contemporary China.

Flora Cottage

"HOME·Communal Garden" consists of a set of daily objects, plants and video devices. It is displayed in parallel to the main exhibition in the garden lawn outside the China National Pavilion, together with the assembled architecture "Dou Pavilion" by architect Zhu Jingxiang. The case is a public welfare project that the View Unlimited design team participated in during 2015, which is still in progress.

The public welfare project is an extension of the design team's completed work on the Yangmeizhu Xiejie renovation project. In order to solve problems such as a lack of arbor plants and poor greening of the environment, which existed before the transformation of Yangmeizhu Xiejie, the design team introduced flower beds during the renovations which connected the street and residential houses which were used to plant ornamental flowers and shrubs in order to increase the overall greening of the street. However, all of these measures aimed at beautifying the environment were not supported by the majority of residents. This phenomenon has led us to rethink a series of issues including public space and public order. Instead of simply attributing the causes of these phenomena to the poor education of the people or a lack of public morality, the team conducted a sociological study of the behind the scenes phenomena relating to the social relations between specific social situations and the specific groups of people through on-the-spot investigation and analysis. The results of the research not only aided us in understanding the living conditions of the people at the bottom rungs of the society, but also altered our undeviating concepts of design. Namely, that design should adapt to the subjects of the design rather than introducing change. Moreover, design should not be imposed on unique subjects in a universal form, regardless of how unsightly it may be. This fear and rejection of heterogeneity is precisely the root of homogenization. Based on this idea, the team commenced construction of the "Flora Cottage" experiment in 2015.

The civilian garden, which has been named "Flora Cottage", is located along Nos. 66-76 Yangmeizhu Xiejie, Dachilan District, Beijing. The passageway itself is 60 meters long, and at its narrowest is 1-meter wide. It is shared by five households, which includes permanent residents, temporary residents and migrant workers. It is a typical old neighborhood in Beijing, which is primarily "a compound occupied by many households". It is a complex residential environment composed of various historical structures, recent urban transformation, cultural street construction and other factors.

With regards to building a co-planting space in a dilapidated passageway, the main purpose was not to create a venue for leisure, entertainment and appreciating flowers, but rather to establish a platform for communication and sharing amongst the five households which are of different backgrounds and origins, neighboring and yet scarcely associating with each other. During the preliminary investigation of the project, the designers found that although the living environments of the five households were very poor, in reality there were also various contradictions and conflicts of interest. As a result, the only common hobby between the five households was growing flowers and vegetables. That is to say, planting was the "greatest common denominator" amongst these inhabitants, which was the intermediary through which they could interact with each other.

Planting and applying greenery are the core elements of landscape architecture design, which are closely related to human visual perception, environment, and ecology in addition to other aspects. In the survey of this public welfare project, designers discovered another social function of planting. In particular, they found that the spontaneous individual behavior of planting can be guided and transformed into a spontaneous public behavior; as long as there is a shared space, there is the possibility of forming a spontaneous community. Through this process, the designer's role also shifts from the philanthropist who is "designing for civilians", and morphs into the guide and participant who is "building for civilians". This transformation of roles implies an equal relationship between the designer and the subject of the designer's service. It is also a response to curator Aravena's statement that "architecture as a shortcut to wards equality".

For a project aiming at sociological significance, the exhibit itself has become insignificant. More importantly, the exhibit lets visitors know what is happening in the original location of the project. The set of installations uses planting containers – pots, pans, foam boxes, and abandoned building materials which were collected by designers in the old urban areas of Beijing. These are the most common "free" containers used by the urban underclass for the purposes of planting, which are recyclable waste in the sense of "environmental protection". Moreover the "display case", on which these containers and plants are placed, are derived from wooden boxes originally used for packaging and transportation.

Five sets of video players are set up for this project. The video content was recorded by designers the week before the exhibition, and includes unedited clips from the life of the five households as well as the real-time scenes of "Flora Cottage". At the same time, there are two sets of videos on the exhibition site which are presenting live streaming. Visitors can see the scenes from the passageway in real time through the player, whilst at the same time the residents in Beijing can see the exhibition through live streaming as well (including the opening ceremony of the China Pavilion). The aim of this is to display the presentation in a manner more similar to a report, which is in the process of implementation and not yet a finished design work. The goal of the exhibition is not to tell a story in an artistic way, by applying editing, special effects and other documentary reproduction techniques, but rather to let visitors know the unadulterated reality of what is happening here.

The Venice Biennale China Pavilion, known as the virgin's garden Garden, features a beautiful environment with soft green grass carpeting the ground on the exhibition site. When further taking into account the flourishing plants and flowers planted amidst the exhibition on display, the first impression of the project may be that it is a garden design. However the live stream video of "Flora Cottage" is in fact a narrow space of less than 10 square meters, featuring a variety of vegetables and fruits and a few ornamental plants which are planted in some foam boxes, paint barrels and other waste collected by residents. Due to this strong contrast, the special significance of the project in the current Chinese context, and the understanding thereof, is also made more accessible to the audience. People may realize that the collection of objects and plants displayed in front of them is not a garden in people's imagination, nor is it a design work, nor a work of art, but rather a report from the scene.

Daily Tao

Daily Tao. Daily use is essentially the habits of everyday life and is an externalization of everyday life. "Tao" is the truth, the meaning of life, and the ultimate goal. This is not only an integral part of traditional Chinese philosophy, but also the fundamental belief in common people's life.

We always adhered to this idea throughout the process of participating in the project of rebuilding Yangmeizhu Xiejie with Mr. Liang Jingyu. However, the logic of reality also makes the entire reform plan full of contradictions and paradoxes. This is a government-led commercial project operated by investment companies, including the relocation of residents, the demolition of illegal buildings, street paving and facade renovation. This involves the problem of capital management and operations. According to the logic of capital: efficiency and value appreciation are the core purposes. Thus, how can the daily life of residents living in this street produce economic benefit? How is it possible to add value? The answer seems to be limited to two options: to relocate, or to change the way the residents live in order to meet the need for capital appreciation. Essentially, daily life is no longer for the purpose of oneself, but is a means to achieve the purposes of others. Under these conditions, daily life is simply a re-coded exhibition of "Hutong Life". If there is no daily life anymore, how can the concept of 'daily use' still exist? As one of the residents mentioned, "This street has become a cultural street now, and has nothing to do with my life."

Although we are well aware of the fact that capital is invincible in the modern world, nevertheless, throughout the process of the street renovation we still adhered to our original intention - to design for civilians and make the renovation project suitable for the daily life of local residents to the fullest extent possible. If we use the metaphor of the "frontline"to describe this environment, then the entirety of our actions have resulted in a society of spectacle constructed from the conflict between capital and daily life.

What does "frontline" mean in relation to the special situation of China? Perhaps the meaning may be found in Aravena's curatorial statement: "More and more people in the planet are in search for a decent place to live and the conditions to achieve it are becoming tougher and tougher by the hour. Any attempt to go beyond business as usual encounters huge resistance in the inertia of reality and any effort to tackle relevant issues has to overcome the increasing complexity of the world."

As a public welfare project, the "Flora Cottage" should have been completed in October 2015. The original intended design for this area was much larger than the present small space, which is less than 10 square meters. In light of this, one can imagine all of the various kinds of resistance our team faced. As a social experimental project, many external uncertainties are difficult for designers to control. Whether or not the results of the project can achieve the originally intended goals requires long-term observation and constant adjustment. However, no matter whatever the result, the project has been a constructive attempt which is outside of the scope of design.

2016.6

安住 · 杨梅竹斜街改造纪实与背后的思考

主创作者

The main designers

第二部分　展览传播

主创作者于威尼斯展品前合影（由左至右为：童岩、谢晓英、黄海涛、瞿志）
The main designers takes a group photo in front of the Venice exhibition (from left to right: Tong Yan, Xie Xiaoying, Huang Haitao, Qu Zhi)

童岩
Tong Yan

Born in 1962, Beijing
Professor of Fine Arts, Renmin University of China

毕业于北京师范大学及比利时安特卫普皇家艺术学院（Royal Academy of Fine Arts）。现任职于中国人民大学艺术学院。作为无界景观团队的设计顾问，参与了多个项目的策划与设计。近年来，他更多地侧重于公共空间、公共艺术与公众生活等方面的理论研究。
工作生活于北京。

谢晓英
Xie Xiaoying

Born in 1964, Beijing,
Urban Designer, Landscape Architect.

毕业于北京林业大学，曾在 Wageningen Agricultural University, Amsterdam Academy of Architecture, Berlage Architecture Institute（Master Class）学习了区域规划、城市设计、建筑设计及景观设计等课程，形成跨领域的设计思想。
2004 年成立无界景观工作室，任主持设计师。
工作生活于北京。

黄海涛

Huang Haitao

Born in 1963, Beijing, Artist

艺术家，无界景观设计团队艺术总监。毕业于中央戏剧学院及比利时根特圣卢卡斯（Sint-Lucas）高等美术学院（Hoger Sint-Lucas Instituut voor Beeldende Kunst Gent），毕业后于布鲁塞尔举办个人画展。曾赴洛杉矶学习计算机艺术并回国从事计算机艺术创作。2009年起多次在国内的西藏、南方以及非洲参加公益建造实践以及少数民族非物质文化调研、采集与策展。近年与建筑师、景观设计师进行跨界合作以及从事个人艺术创作。

工作生活于北京。

瞿志

Qu Zhi

Born in 1965, Hunan
Professor of Landscape Architecture, Beijing Forestry University

园林设计师，无界景观团队工程设计总监，毕业于北京林业大学，现任该校副教授。从事园林教学、科研和工程实践，曾作为公派访问学者在美国德克萨斯（TEXAS A&M）大学建筑学院研习。曾多次赴非洲进行生态景观项目研究与实践。关注科学、艺术、生态等多领域的融合和地域特征的传承。

工作生活于北京。

安住 · 杨梅竹斜街改造纪实与背后的思考

团队设计的 POSTCARD（部分）
Some of the Postcards designed by the team

第二部分　展览传播

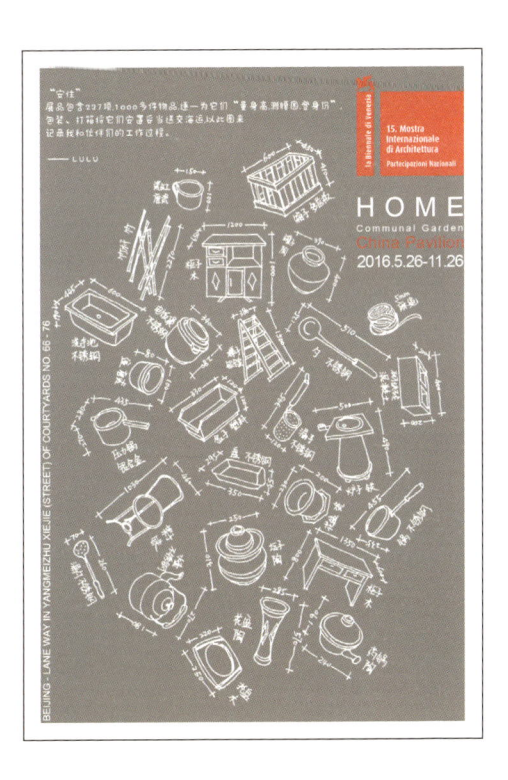

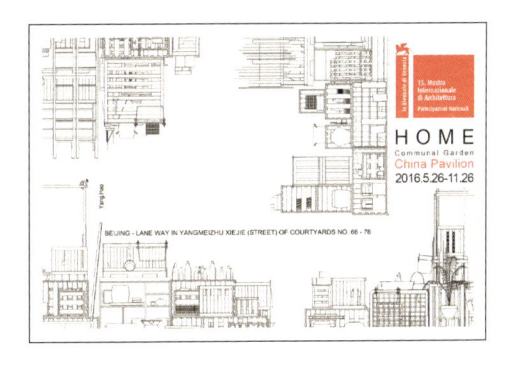

215

2017 年居民种植展播种仪式
2017 seed planting ceremony

第二部分　展览传播

二、杨梅竹花草堂 2016
Yangmeizhu Xiejie Hutong Flora Cottage 2016
——日常生活的景观
Daily Tao VS Spectacle

策展人：童岩
Curator: Tong Yan

2016年10月北京国际设计周期间，无界景观设计团队与杨梅竹斜街居民一起举办了"杨梅竹花草堂2016"活动。活动以"平民花园——日常生活"VS"景观设计"为主题，没有成本、没有设计、没有"创新"、没有刻意的视觉营造、没有环保主义的说教与劝导，只是展现普通居民的种植经验以及他们日常生活中的智慧。

正如参展的段大爷所说："咱们办花展，就是为了普及、推广，把大家的兴趣培养起来。大伙都种花，都参与，将来咱们就不是在小院办花展了，而是在杨梅竹街上办花展了，这条街的绿化就起来了"。

此次设计周临时展览，是继我们在 2015 年设计周在杨梅竹斜街 66—76 号院策划的夹道"花草堂"活动和 2016 年 5 月份杨梅竹设计月活动的延伸和发酵，是探讨设计参与社区营造的进一步尝试。

展览分为两部分，第一部分是我们以 66—76 号院夹道为原型的、目前正在第 15 届威尼斯国际建筑双年展展览中的作品"安住·平民花园"的图片展示；第二部分是杨梅竹斜街居民日常种植展览。我们邀请杨梅竹斜街热爱种植的居民参与展览，展品由本地居民提供，将夹道"花草堂"活动拓展到整条杨梅竹斜街。

展览期间，我们通过花草展示、举办讲座、互赠种子、售卖明信片等方式鼓励居民之间、居民和参观者之间的互动。让胡同百姓展示生活中的智慧、分享种植经验、促进邻里间的交往，让设计周不再只是游客的设计周，让设计回归日常生活。

第二部分　展览传播

花草堂 2016 展览中前来参观的游客
Visitors observing the "Floral Cottage" at the 2016 Expo

安住 · 杨梅竹斜街改造纪实与背后的思考

NO COST, NO DESIGN, NO "INNOVATION".
YEAR AFTER YEAR, PLANTING IS JUST A PART OF THEIR
DAILY ACTIVITIES.

WITHOUT DELIBERATE VISUAL FURBISHMENT,
THIS IS WHAT FLOWS IN THE BLOOD OF LIVES OF THE
COMMONALITIES: FORM AND CONTENT AS ONE.

IT SEEMS THAT NO ONE NEEDS A LESSON FROM
ANY ENVIRONMENTALISTS ABOUT HOW ABANDONED
INDUSTRIAL PRODUCTS COULD BE RECYCLED,
OR HOW ENERGY COULD BE SAVED WITH LOCAL
PRESERVATION METHODS...
PEOPLE JUST USE A BIT OF DAILY LIFE WISDOM TO
REFLECT THEIR OWN AESTHETICS IN LIVING.

The flowers and plants were self-planted by residents in Yangmeizhu Xiejie and the nearby streets, they have adapted to the crowded neighborhood situation and cleverly constructed small personal spaces with very limited costs. Despite the substandard living environment, the local residents find gaps in their small living spaces where sunlight could penetrate through to plant vegetables and flowers using abandoned debris and waste containers. The residents, majority who live near the bottom of society, take care of the harvesting of vegetables and seasonal blossoms in an almost ritual way to comfort their minds. These daily practices seem to be apparently opposite to what developmentalism directs, yet we might perceive these moves as an alternative reference system under current consumerism-dominated world.

花草堂 2016 展览现场
"Flora Cottage" at the 2016 Expo

安住 · 杨梅竹斜街改造纪实与背后的思考

赵阿姨（赵琴，65岁）高叔（高魁，66岁）
杨梅竹斜街113号
退休前均为北冰洋汽水厂职工，北京人
2000年搬至杨梅竹斜街

"在红桥儿住的时候我种月季，院儿里没地儿种房顶上，爬梯子一天浇两回水。
街坊都说：'那是你的花啊，真漂亮。'"

Ms. Zhao (65), Mr. Gao (66). Live in No. 113 Yangmeizhu Xiejie, from Beijing.
Now retired, they used to both work for Beibingyang Soda Factory.
They moved to Yangmeizhu Xiejie in 2000.

"I used to plant Chinese roses when I lived in Hongqiao, there was not enough space in the courtyard
so I moved the flower to the roof. I would have to climb the ladder twice a day to water the flower.
My neighbor would tell me how beautiful my flowers were."

第二部分　展览传播

孙奶奶（孙淑华）
杨梅竹斜街 119 号
86 岁，北京人
退休前在国泰照相馆工作
在杨梅竹斜街居住二十余年

"我每天早上起来，两小时，连归置带浇水，就当运动了。"

Grandma Sun, 86, lives in No. 119 Yangmeizhu Xiejie, from Beijing.
Now retired, Sun used to work in Guotai Photo Studio.
She lived in Yangmeizhu Xiejie for over 20 years.

"Every morning right after getting up from bed,
I would spend two hours in tidying up and watering plants. It's like morning exercise."

安住 · 杨梅竹斜街改造纪实与背后的思考

魏家（魏兰涛及家人）
杨梅竹斜街 76 号
搬来杨梅竹二十多年，祖孙三代居住于此
魏叔和老爷子都爱种花种菜，并将果实分享给邻里，
女儿在这里出生长大，在附近的师大附中上学。

"以前我工作挺忙的，现在不忙了，平时爱养养花。"

The Wei Family, lives in No. 76 Yangmeizhu Xiejie,
three generations lived here for over 20 years,
Mr. Wei and his father both love planting vegetables and often shares them with the neighbors.
Wei's daughter was born here and now goes to the High School Affiliated to Beijing Normal University nearby.

"I used to be busy, now I have plenty of free time to plant flowers."

第二部分　展览传播

东大妈（东雪梅）
杨梅竹斜街 72 号
62 岁，北京人
退休前在北京第二制药厂工作
在杨梅竹斜街居住 21 年

"原来我种的少，现在越来越多。播种也有讲究，种子小少盖土，种好之后土得拍瓷实了。"

Ms. Dong, 62, living at No. 72 Yangmeizhu Xiejie,from Beijing.
Now retired, Ms. Dong perviously worked at Beijing No. 2 Pharmaceutical Factory.
She has lived in Yangmeizhu Xiejie for over 21 years.

"I used to plant less, and now it's getting more and more everyday. You really need to take special care when sowing, if the seeds are small you should put on less soil on top, and pat the soil firmly when done."

段大爷（段宝玺）
百合园胡同 10 号，
68 岁，胡同退休手艺人，陕西人
移居北京百合园胡同 20 年．
原中医正骨大夫、木工、瓦工、冶炼工，爱好根雕盆景制作、书画

"咱们办花展，就是为了普及、推广，把大家的兴趣培养起来，大伙都种花，都参与，将来咱们就不是在小院办花展了，在杨梅竹街上办花展了，这条街的绿化就起来了。"

Mr. Duan (Baoxi Duan), 68, living at No. 10 Baiheyuan Hutong, from Shaanxi.Migrated to Beijing Baiheyuan Hutong 20 years ago, Mr. Duan used to be a practitioner of traditional Chinese medicine specializing in osteopathic manipulation. He was also a retired Hutong craftsman, a carpenter, a bricklayer and a smelter. Mr. Duan is interested in root-carving, bonsai-making and traditional painting and calligraphy.

"The purpose of us having a flower show is to promote and encourage this lifestyle to everyone, so that more people will be interested in planting and participate in planting and gradually there will be larger shows. We will develop the flower show from this little courtyard to Yangmeizhu Xiejie, and eventually the entire street will have better greening."

第二部分　展览传播

为居民举办种植课堂
Planting classes for residents

安住·杨梅竹斜街改造纪实与背后的思考

参展的邻居们合影（摄影：张元）
Group photo taken with the neighboring exhibitors
(Photo by Zhang Yuan)

第二部分　展览传播

由左至右：魏兰涛、高魁、赵琴、段宝玺、孙淑华、东雪梅

三、杨梅竹花草堂 2017
Yangmeizhu Xiejie Hutong Flora Cottage 2017

—— "众"瓜得瓜"众"豆得豆
Together we sow, Together we harvest

策展人：童岩
Curator: Tong Yan

在 2015 年和 2016 年连续两年举办"胡同花草堂"种植展的基础上，2017 年北京国际设计周期间，无界景观设计团队继续带来了"杨梅竹花草堂 2017"主题活动。本次展览以胡同中最常见的、本地居民用作种植器皿的泡沫包装箱为载体，配以他们日常种植的图片和所种植物，旨在展示胡同居民生活的日常景象——年复一年，日复一日；种瓜得瓜，种豆得豆。展览结束后，展示用的容器和种植土全部留给居民继续种植。

Continue from the successful planting exhibition of Hutong Flora Cottage in 2015 and 2016, in 2017 Beijing Design Week, the design team from View Unlimited had curated another themed planting exhibition: "Yangmeizhu Flora Cottage 2017".The exhibition features the foam boxes which are most commonly used by the Hutong residents as planting containers. Along with the images showing their daily planting activities, the installation showcases the repeatitive, mediocre but somehow peaceful and poetic Hutong lifestyles: as you sow, so shall you reap.After the exhibition, the vessels and planting soil were left to the residents for future planting.

第二部分　展览传播

安住·杨梅竹斜街改造纪实与背后的思考

113号（2014）
No. 113(2014)

第二部分　展览传播

2017 年北京国际设计周 113 号
2017 Beijing Design Week No. 113

安住·杨梅竹斜街改造纪实与背后的思考

98 号（2015）
No. 98(2015)

第二部分　展览传播

2017 年北京国际设计周 98 号
2017 Beijing Design Week No. 98

在展览开幕时，参加种植展的老邻居面对记者的镜头，难掩兴奋与自豪，在一小时内换了三身衣服
During the exhibition opening, the elderly neighbor participating in the planting exhibition couldn't hide their excitement and pride in front of the camera, changing outfits three times within one hour

第一部分　展览传播

2017 年北京国际设计周 93 号
2017 Beijing Design Week No. 93

参加种植展的老邻居夫妇（左二、三）为游客们介绍自家种的丝瓜
The elderly neighbor couple participating in the planting exhibition (second and third pictures from the left) introduce the towel gourds they grow to the visitors

2017 年北京国际设计周 45 号
2017 Beijing Design Week No. 45

安住 · 杨梅竹斜街改造纪实与背后的思考

塑料泡沫展箱

展览期间

放入种植土作为基座

植物

种植土

展览后展品利用

种植箱

第二部分　展览传播

展览后展品被居民利用种植植物
After the exhibition, the exhibited items are replanted by the residents

四、三岁，胡同花草堂
Three-Year-Old, Hutong Flora Cottage
——杨梅竹斜街66—76号夹道社区营造项目三年展
Three Years of Revitalizing the Alley 66-76 Yangmeizhu Xiejie

策展人：黄海涛
Curator: Huang Haitao

截止到2017年北京国际设计周举办，无界景观设计团队对杨梅竹斜街66—76号夹道进行社区营造已经进行了三年有余。为此，我们策划了"三岁，胡同花草堂"主题展出，回归北京大栅栏片区杨梅竹斜街66—76号夹道社区营造的缘起及我们为此做出的努力。通过现场种植、早期方案设计展示以及三年间的图片实物文献组成现场装置等形式，展现花草堂的前生来世。

胡同花草堂是一个具有一定实验性的项目，不求速效，也不会速朽。生活中任何微小而切实的改变，对于居民都意义重大。北京胡同居民的成分早已在市场化的过程中改变。此胡同已不是彼胡同，如何为现有的胡同保存生机，是一个有待持续关注的课题。在现有的胡同中保存曾经有过的丰厚的文化意蕴，才是真正的挑战。

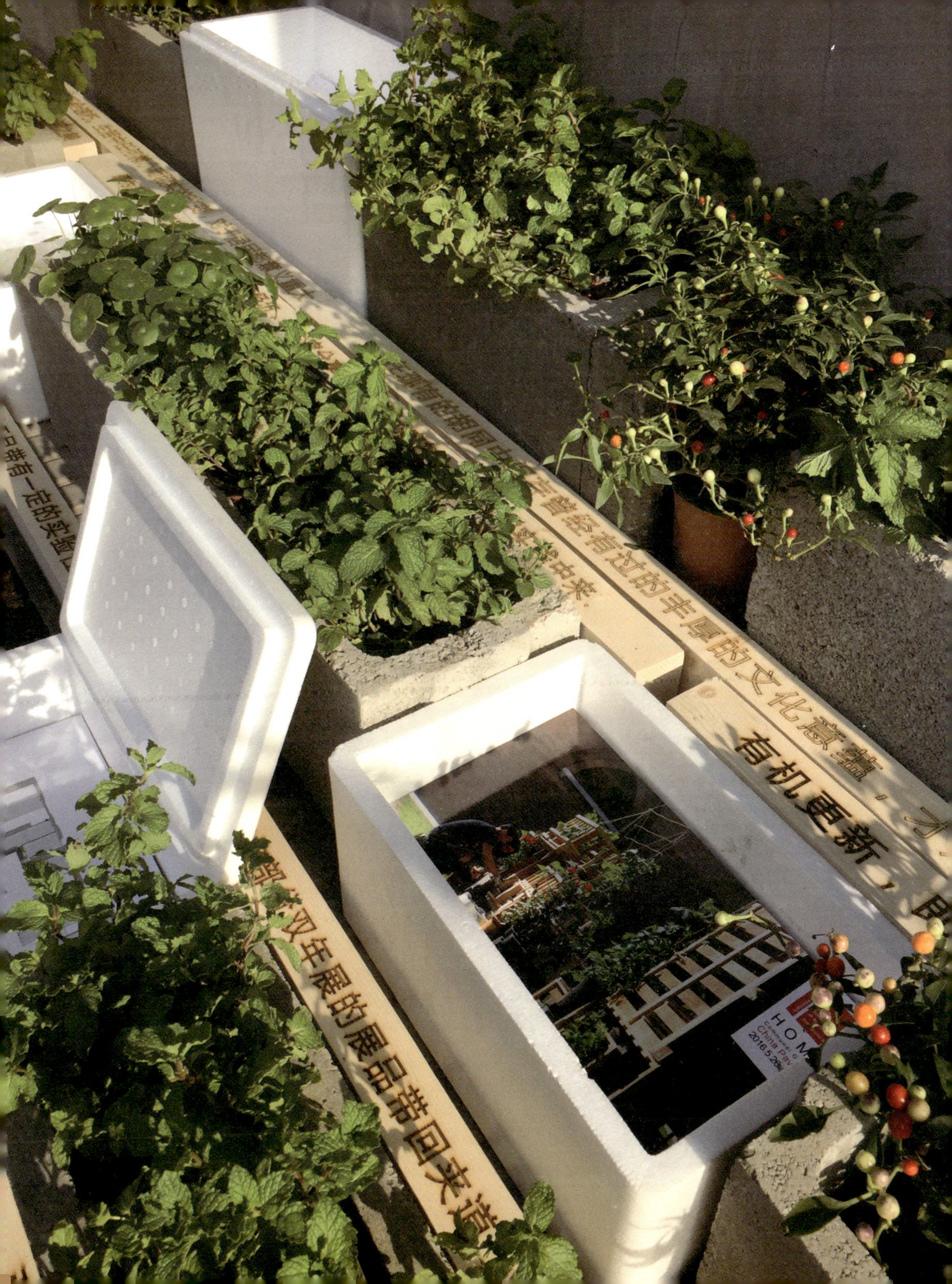
曾经有过的丰厚的文化意蕴，
有机更新，
双年展的展品带回来造
带有一定的实验

安住 · 杨梅竹斜街改造纪实与背后的思考

As of the Beijing Design Week held in 2017, the View Unlimited Landscape Architects design team has been involved in community revitalization efforts for over three year in the Alley 66-76 Yangmeizhu Xiejie. To showcase the origins of the community revitalization project in this area of Dashilan, we have planned a themed exhibition titled "Three Years Old, the Hutong Flora Cottage". Through on-site planting, displays of early design schemes, and a collection of images and physical documents from the past three years, we aim to present the past and future of Flora Cottage.

Hutong Flora Cottage is an experimental project that prioritizes long-term impact over immediate results. Any tiny and tangible change in one's life is of great importance with regards to residents. Through the process of marketization the composition of the residents within the Beijing hutongs has already experienced great changes. This alley is no longer the hutong of yore. The question of how to preserve the way of life in the existing hutongs is an ongoing one of vital importance. The true challenge is to preserve the rich cultural connotations that presently exist in the hutongs.

第二部分　展览传播

2017 年北京国际设计周"三岁，胡同花草堂——杨梅竹斜街 66—76 号夹道社区营造项目三年展"
2017 Beijing Design Week
"Three-year-old, Hutong Flora Cottage - Three Years of Revitalizing the Alley 66-76 Yangmeizhu Xiejie".

五、软组织·胡同中的即时健身系统
Soft Tissues: Spontaneous Exercise System in Hutongs

策展人：谢晓英
Curator: Xie Xiaoying

狭窄细碎的胡同街巷构成的空间可以成为随时随地的健身场所。为此，我们设计了一套简单有效的、徒手就可以进行的练习，以展示牌、张贴画等形式，营造了一种软性的、安全的、温和的、能够随时随地健身的引导系统，使居民能够利用身边的日常构筑物进行最基础的身体拓展训练，让科学健身的新方法进入老城最深处的细碎空间，达到城市的"末梢神经"，辐射没有健身意识的人群。

The spatial quality of Hutong is narrow and fragmented, and there are spaces to be utilized for spontaneous exercises.For this purpose, we have designed a set of simple and effective exercises that can be performed without any equipment. Through the use of signs, posters, and other visual aids, we have created a gentle and safe fitness system. This system directs the local residents to use existing structures for basic body exercises. It acts as a silent instructor, providing scientific exercise instructions for those who live in the nerve ending of the city.

第二部分　展览传播

安住·杨梅竹斜街改造纪实与背后的思考

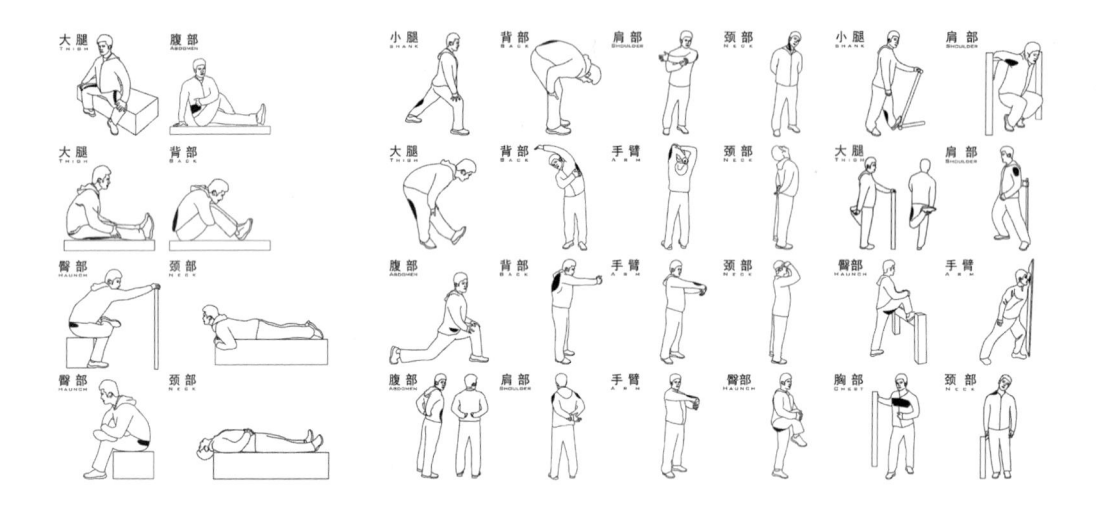

第二部分　展览传播

无器械健身系统不仅让住在杨梅竹斜街胡同里的居民能跟着健身标识运动，也能吸引来到这里的游客运动打卡。

The accessible fitness system not only allows residents living in the Hutongs of Yangmeizhu Xiejie to exercise following the fitness signs but also attracts tourists visiting here to exercise and check-in.

一名法国建筑师正在跟着无器械健身标牌做运动

Residents follow guides for fitness exercises

六、杨梅竹斜街小气候监测数据交互展
Exhibition: Interactive Display of Microclimate Monitoring Data on Yangmeizhu Xiejie

策展人：杨鑫、张琦
Curator: Yang Xin, Zhang Qi

城市双修是转变城市发展方式有效手段，此次展览旨在从老城区小气候环境研究的视角，探索老城区人居环境改善与公共空间修补的有效方式。展览将杨梅竹斜街小气候环境的变化以大数据分析与实时监测的形式直观展现出来，探索老城区小气候环境的现状特征及改善措施，同时提升人们对气候变化的关注，呼吁老城区小气候适应性改造设计的提出。

Urban dual restoration, namely, ecological restoration and urban repair, is an effective approach to transforming urban development methods. This exhibition aims to explore effective approaches to improving the living environment and repairing public spaces in the old city from the perspective of microclimate research.The exhibition visually presents the changes in the microclimate environment of Yangmeizhu Xiejie through data analysis and real-time monitoring, exploring the current characteristics and improvement measures of the microclimate environment in the old city. Additionally, it raises awareness about climate change and advocates for adaptive redesign of the microclimate in the old city.

第二部分　展览传播

位于胡同微公园的监测现场照片
Photos from live streaming monitoring at the Hutong Mini Park

位于胡同微公园的监测现场照片
Photos from live streaming monitoring at the Hutong Mini Park

第二部分　展览传播

七、杨梅竹花草堂 2018
Yangmeizhu Xiejie Hutong Flora Cottage 2018
——繁华落尽终归平常
When All Comes Down to Ordinary

策展人：谢晓英、童岩
Curator: Xie Xiaoying, Tong Yan

自 2012 年北京大栅栏片区杨梅竹斜街环境更新项目开始以来，这条街道在过去的六年里发生了很大的变化。一方面，不断有新的餐厅、咖啡馆、文艺小店开张或关门，也许这正是文化创意产业的活力所在；另一方面，常驻于此的居民的生活却是一如既往，少有可察觉到的明显变化。去年北京设计周期间，我们曾经策划过将街道居民自家种的丝瓜、葫芦、豆角、辣椒等通过包装设计实现个人或家庭种植的品牌化的活动，目的是让文化创意活动也能够惠及街道里的普通居民，但获得的反馈却是：他们都不愿意参与这项活动。

多数居民表示他们的种植行为纯属生活习惯，从没想过也不愿参与什么创意或交易的事。从某种角度看，这样的结果也许可以归结为老北京人的惰性，但他们随遇而安的自在性却是一种与现代文明平行的、可互为参照的、实实在在的文化存在。繁华落尽日，终归平常时。在设计、时尚的浮光掠影表象下，老北京人那种在过往岁月中积淀下来的日常生活习性将延绵不绝，代代相承。

Six years have passed since the beginning of the reconstruction project of Yangmeizhu Xiejie, and the street had changed drastically since then. Businesses came and went, many creative industries such as new restaurants, cafes and arty grocery stores had made attempts in this street. On the other hand, the local residents' life styles remained unchanged, or at least not obvious enough to be observed directly. During the Beijing Design Week last year our team had tried to convinced the local neighborhoods to participate into the creative cultural activities, and we suggested to design home-made packages for their own plants and vegetables (such as zucchini, gourds, green beans, chillies etc). Interestingly, none of them wanted to be part of the "home-made branding" activity.

A majority of the residents told us that the reason for planting was just a daily life habit, they have never thought of, nor that they would like to, participate in any creative-involved or trading businesses.

One could argue that such mentality is due to the typical "sluggishness" of the Beijingnese, but it is such inertia ease that had been existing parallel to the fast-modern culture. At the end of the day, when all designs, fashion and trends are stripped away, it all comes down to ordinary. The local Beijingnese would keep their daily lifestyles just as what they have done casually for centuries, and for many generations in the future.

113 号居民在自家门前照料花草
Residents of No. 113 caring for flowers and plants in front of their own doors

2018 年北京国际设计周 113 号
2018 Beijing Design Week No. 113

安住 · 杨梅竹斜街改造纪实与背后的思考

2018 年北京国际设计周 45 号
2018 Beijing Design Week No. 45

八、"花草堂"的延伸
Flora Cottage Expansion

2016 北京副中心行政办公区镜河水系"HOME·安住"公共艺术装置
2016 "HOME" Public Art Installation in the Administrative Office Area of the Jing River System, Beijing Sub-Center

2016 年，设计团队展开了对北京城市副中心行政办公区镜河水系的景观设计工作。镜河与通惠河、运潮减河、北运河承接了北京古城的水韵，延伸了古都文脉，是副中心"北京气质"的载体。在镜河河岸游步道，我们与北方石匠共同创作了一件长约 1.6 千米、宽 30 厘米的大型当代艺术装置作品"HOME·安住"，这是"安住·平民花园.HOME.COMMUNAL GARDEN"的延伸作品。该作品由镌刻着千余条北京胡同名字的新旧石料，相互以榫卯的方式契合衔接，在满足镜河河畔健身慢跑径封边石（道牙）功能的同时，兼具承载北京数百年间城市发展线索的文化功能。在装置作品一侧，我们将功能性的挡土墙设计成毛石挡墙，在砌石缝隙填充土壤后，带领居民在挡土墙上撒播花籽，种植花草，与装置作品构成一条能够储存北京民众共同记忆的情感线。

安住·杨梅竹斜街改造纪实与背后的思考

2016年，设计团队开始进行北京城市副中心行政办公区镜河水系的景观设计工作。镜河与通惠河、运潮减河、北运河承接了北京古城的水韵，延伸了古都文脉。从北京杨梅竹斜街（A）到北京副中心行政办公区（B），城市公共空间设计在满足使用功能的同时，兼具承载北京数百年间城市发展线索的文化功能。

In 2016, the View Unlimited design team embarked on landscape design work for the Jing River system in the administrative office area of the Beijing Sub-Center. The Jing River, along with the Tonghui River, the Yunchaojian River and the Beiyun River, carries the water's charm of ancient Beijing, extending the cultural heritage of the ancient capital. From Beijing Yangmeizhu Xiejie (A) to the administrative office area of the Beijing Sub-Center (B), the design of urban public spaces fulfills functional requirements while also has the cultural function of carrying the clues of Beijing's urban development over the past few hundred years.

第二部分　展览传播

安住．平民花园　············双年展砌块
HOME.COMMUNAL GARDEN
La Biennale Di Venezia 2016　············铜板刻字

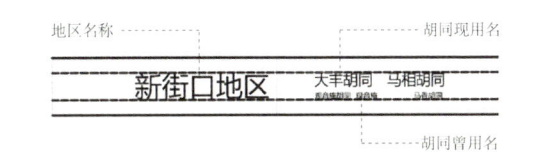

地区名称············　　　　　············胡同现用名
新街口地区　大丰胡同　马相胡同
············胡同曾用名

"安住·平民花园"参展物件被浇筑在混凝土砌块内，砌筑于封边石中 The exhibits of "HOME.COMMUNAL GARDEN" are embedded within concrete blocks and incorporated into the edge stones.

"HOME·安住"是"安住·平民花园.HOME.COMMUNAL GARDEN"的延伸作品，作品在记载了胡同名称的同时，也记录了部分胡同名称在数百年间的演化，储存了丰富的文化信息。（2021）
"HOME" is an extension of "HOME - COMMUNAL GARDEN" installation. Alongside capturing alley names, it also traces the evolution of these names over centuries, preserving a wealth of cultural heritage.(2021)

安住 · 杨梅竹斜街改造纪实与背后的思考

In 2016, the View Unlimited design team embarked on landscape design work for the Jing River system in the administrative office area of the Beijing Sub-Center. The Jing River, along with the Tonghui River, Yunchaojian River and the Beiyun River, carries the water's charm of ancient Beijing, extending the cultural heritage of the ancient capital and serving as a conveyor of the "spirit of Beijing" in the sub-center. Along the bank trail of the Jing River, we collaborated with northern stonemasons to create a large-scale contemporary art installation titled "HOME" spanning approximately 1.6 kilometers in length and 30 centimeters in width. This installation was an extension of the comprehensive installation artwork "HOME - COMMUNAL GARDEN", which was selected for the Chinese Pavilion at the 15th International Architecture Exhibition in 2016.

The artwork is comprised of old and new stone material engraved with over a thousand names of Beijing hutongs , fitting together in a dovetail manner. While serving as edge stones (curbs) for the fitness and jogging paths along the Jing River, the artwork also serves a cultural function, hinting at the urban development of Beijing over the centuries. On one side of the installation, functional retaining walls were designed as drystone retaining walls. After filling the stone joints with soil, residents were encouraged to sow flower seeds and plant flowers and plants on the retaining walls, forming an emotional connection with the installation. This created a thread of shared memories for the people of Beijing.

2020 埃塞俄比亚谢格尔公园友谊广场"播种明天"公共艺术活动

2020 "Sowing Tomorrow" Public Art Event at Friendship Square in Sheger Park, Ethiopia

2019 年 4 月,设计团队承接了埃塞俄比亚河岸绿色发展项目,该项目的核心组成部分友谊广场(谢格尔公园)位于亚的斯亚贝巴市中心,是具有庄严的国家政治属性以及埃塞民族、历史文化特色的城市中心广场。2020 年 1 月 1 日,埃塞俄比亚谢格尔公园"都市花草堂"计划启动,设计团队向当地居民传授花草种植技术,带领当地居民在长达千米的挡土墙上种植花草,开启了在非洲的以花草种植为媒介,让当地民众愉快交流的公共艺术活动。

In April 2019, our design team undertook the Ethiopia Riverbank Green Development Project. The core component of this project, Friendship Square (Sheger Park), is located in the heart of Addis Ababa, Ethiopia's capital, and serves as a dignified national political venue with distinct Ethiopian national and historical cultural characteristics. On January 1, 2020, the Urban Flora Cottage project was launched at Sheger Park. The design team imparted flower and plant cultivation techniques to local residents, guiding them in planting flowers and plants on a retaining wall stretching for kilometers. This initiative marked the beginning of public art activities in Africa that promote joyful interactions among local people through flower and plant cultivation.

安住·杨梅竹斜街改造纪实与背后的思考

2020年1月1日，埃塞俄比亚谢格尔公园友谊广场"都市花草堂"启动，举办了"播种明天"公共艺术活动，活动名称引用了印度谚语"今天的种子就是明天的花朵"。埃塞俄比亚总理阿比亲临现场，在挡土墙的石缝中播种花草。

On January 1, 2020, the Urban Flora Cottage initiative was launched at Friendship Square in Sheger Park, Ethiopia, with the "Sowing Tomorrow" public art event. Inspired by an Indian proverb, the event's name reflects the idea that "All the flowers of all the tomorrows are in the seeds of today".
Ethiopian Prime Minister Abiy Ahmed joined the occasion, planting flowers and plants in the stone crevices of the retaining wall.

第二部分　展览传播

埃塞的居民热情参与互动，目前这座"花墙"已经初具规模（摄于 2020 年和 2022 年）
Local residents enthusiastically engaged in the interactive event, and now, this "flower wall" is taking shape (Photos taken in 2020 and 2022)

265

2021 "杨梅竹胡同花草堂"与"都市花草堂"民间文化交流活动
2021: Cultural Exchange Activities between the "Yangmeizhu Hutong Flora Cottage" and "Urban Flora Cottage"

2021年9月埃塞俄比亚新年庆典期间，设计团队策划举办了杨梅竹斜街"胡同花草堂"与谢格尔公园"都市花草堂"的现场连线活动，发挥"一带一路"的民间使者的作用，促进民心相通，让两地普通百姓互相了解，建立友好连接。

During the Ethiopian New Year celebrations in September 2021, our design team organized a live event connecting the Hutong Flora Cottage in Yangmeizhu Xiejie with the Urban Flora Cottage in Sheger Park. Leveraging the grassroots role of the Belt and Road Initiative, we aimed to foster connections between people, promote mutual understanding, and build friendly ties between the two locations.

第二部分　展览传播

2021年9月，杨梅竹斜街"胡同花草堂"与埃塞俄比亚谢格尔公园友谊广场"都市花草堂"现场连线活动
In September 2021, a live online event was held between the Hutong Flora Cottage in Yangmeizhu Xiejie and the Urban Flora Cottage at Friendship Square in Sheger Park, Ethiopia

267

在 2017 年杨梅竹花草堂《软组织·即时健身系统》展览结束后，设计师们并未停止对便捷健身理念的推广与实践，他们积极响应居民需求，持续鼓励胡同居民开展灵活多样的日常健身活动。

2018 年，设计团队将即时健身系统的理念植入北京副中心镜河水系的景观设计之中，设计了一条全长约为 3.6 千米，与河岸绿道紧密结合的健身路径，提供了 20 处无器械健身场地，让健身不再局限于室内场馆，而是融入风景之中，激发人们的健身热情，增加相遇交流的可能性，从而促进了邻里间良好社会关系的建立与发展。

在 2020 年全球疫情的大背景下，设计团队因地制宜，将即时健身系统引入谢格尔公园，构建了一个与日常生活环境紧密相连的健身空间，鼓励居民拥抱健康的生活习惯，增强健康管理意识，通过轻松愉快的户外活动体验，全面提升自身身体素质和抵抗健康风险的能力，同时也促进了社区内的友好互动和团结精神。

2021 年，设计团队创作了视频作品《抗体·风景融入日常生活——即时健身系统》，该作品展示了我们在北京杨梅竹斜街、北京副中心镜河河岸以及埃塞俄比亚谢格尔公园设计项目中注入的这一"软性、安全、温和"、能够随时随地健身的引导系统。该作品入选第四届中国设计大展公共艺术专题展，并于 2023 年 3 月在中国深圳展出。

第二部分　展览传播

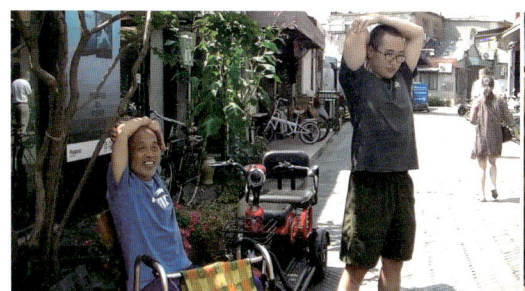

杨梅竹斜街健身活动（2021.9）
Yangmeizhu Xiejie fitness activities

北京城市副中心镜河沿岸健身活动（2021.10）
Fitness activities along the Jing River in Beijing's sub-center

谢格尔公园友谊广场健身活动（2021.9）
Fitness activities in Friendship Square in Sheger Park

安住 · 杨梅竹斜街改造纪实与背后的思考

作品编号 Serial number : 40303

抗体 风景融入日常生活 即时健身系统

设计师/作者 Designer/Creator	谢晓英 黄海涛 周欣萌 王翔 王欣
作品说明 Work description	该组作品取自无界景观工作室完成的位于北京杨梅竹斜街、北京副中心镜河河岸以及埃塞俄比亚首都亚的斯亚贝巴市中心谢格尔公园的设计项目。从老城胡同，到新城河岸，再到城市中央公园，设计团队在城市开放空间注入一种"软性、安全、温和"、能够随时随地健身的引导系统，诱导没有健身意识的人群，形成健康的、积极乐观的生活方式，为促进城市与社会的稳定发展提供一种创新的、可推广的模式。
创作时间 Creation time	2021

270

After wrapping up the "Soft Tissue — Accessible Fitness System" exhibition at Yangmeizhu Flora Cottage in 2017, we didn't stop there. Responding to residents needs, we continued to encourage a variety of daily fitness activities among hutong residents.

In 2018, we brought the concept of an accessible fitness system into the landscape design of the Jing River system in Beijing's sub-center. Our team created a 3.6-kilometer-long fitness path that seamlessly integrated with the riverside greenway, offering 20 equipment-free fitness sites along the route. By merging fitness with the natural beauty of the surroundings, our initiative aimed to inspire a passion for exercise and foster social connections within local communities, breaking away from the constraints of indoor fitness facilities.

Against the backdrop of the global pandemic in 2020, our design team introduced accessible fitness systems to Sheger Park that were specifically tailored to local conditions. This initiative created fitness spaces closely connected to daily life, inspiring residents to adopt healthy lifestyle habits, heighten their awareness of health management, and enhance their physical fitness while minimizing health risks through enjoyable outdoor activities. Moreover, it fostered community interaction and solidarity, promoting a sense of togetherness among residents.

In 2021, we produced the video "Vitality: Landscapes in Daily Life — Accessible Fitness System," showcasing the gentle and safe approach to fitness we implemented in Yangmeizhu Xiejie, the Jing River riverbank in Beijing's sub-center, and Sheger Park in Ethiopia. This work was featured at the 4th China Design Exhibition and Public Art Thematic Exhibition in Shenzhen, China, in March 2023.

设计一种"软性、安全、温和"的城市景观：
Designing a "universal, safe and welcoming" urban landscape:
用流动的善意打开心扉
Open your heart to the power of kindness

在新冠肺炎病毒尚未完全散去的城市，人人都多了一份警觉和自我保护意识，社交距离和私人领地感不可避免地成为公众关注的焦点。在日益繁忙的城市，工作环境以及交互性匮乏的办公、生活空间的双重挤压下，城市陌生人社交与邻里关系也遇到了很大的挑战。如何在保障基本接触安全的前提条件下，提高公共空间的基本社交需求，改善邻里之间的交流状况，倡导城市资源的良性共享，是设计工作者亟待探讨和解决的课题。入选了第4届中国设计大展暨公共艺术专题展的作品《抗体·风景融入日常生活——即时健身系统》便通过研究分析疫情期间的社区连接，从连接美学和关系美学的维度上思考后疫情时期的公共艺术发展。由中国城市建设研究院无界景观设计团队创作完成的这组作品取自无界景观工作室完成的位于北京杨梅竹斜街、北京城市副中心镜河河岸以及埃塞俄比亚首都亚的斯亚贝巴市中心谢格尔公园的设计项目。从老城胡同，到新城河岸，再到城市中央公园，设计团队在城市开放空间注入一种"软性、安全、温和"、能够随时随地健身的引导系统，引导没有健身意识的人群，形成健康的、积极乐观的生活方式，为促进城市与社会的稳定发展提供一种创新的、可推广的模式。设计师希望通过运用城市景观设计的方式,打造城市公共空间中人与人交流的场所,或通过设计为健身等与居民息息相关的生活需求服务,用流动的"善意"重新打开人们久闭的心扉,注入城市社区的各个角落。它以一种方式补充了创造性活动,使来自不同情况和文化背景的人在我们相遇和联系时能够找到相似的体验和情感共鸣。正是在这些深入的交流中,人们开始体会到彼此间的相互陪伴,同时也能参与到他人的思考和创作过程中,渐渐地,我们成为彼此的情感体验者和创造性合作者,共同探索艺术实践和社区关怀的重要作用和文化价值。(文字取自作品说明和第4届中国设计大展公共安全板块主题阐述)

In cities where COVID-19 has not yet fully dissipated, everyone is on high alert and social distancing and a sense of private space have inevitably become a focus of public attention. In an increasingly hectic urban working environment, which suffers from a scarcity of interactive office space and living space, socialization between individuals is increasingly limited. In light of this, design professionals have begun to urgently explore how to enhance the basic social needs of public spaces, improve communication between neighbors, and advocate the sharing of urban resources under the premise of guaranteeing the basic health and safety of each individual. The work "Antibodies - Landscape in Daily Life - Accessible Fitness Systems" was selected for the 4th China Design Exhibition & Public Art Thematic Exhibition. This piece considers the connections that have bound together communities during the epidemic and the development of public art in the post-epidemic period through the lens of connective-relational aesthetics. This set of works, conceived by the View Unlimited Landscape Architects Studio Design Team and China Academy of Urban Planning and Design, was taken from projects completed by View Unlimited Landscape Architects Studio in Beijing's Yangmeizhu Xiejie, the Jing River riverbank in Beijing's urban sub-center, and Sheger Park in downtown Addis Ababa, Ethiopia. From the old city's hutongs to the new city's riverbank, and an urban central park, the design team injected a "universal, safe and welcoming" guided fitness system into urban open spaces, which can inspire individuals with poor health and fitness habits to embrace a healthy, active, and positive lifestyle, and provide an innovative and replicable model to promote the stable development of cities and societies. Designer hopes to use urban landscape design as a tool to create a platform for human communication in urban public spaces, which can serve residents' essential needs, such as health and fitness, and energize people's guarded hearts by injecting the power of "kindness" into every corner of urban communities. In her words, Urban landscape design complements creative activities in a way that allows people from different contexts and cultural backgrounds to find similar experiences and an emotional resonance when we meet and connect. It is in these deeper exchanges that people begin to experience each other's companionship, while also being able to engage with the thoughts and creative processes of others. Gradually, we become emotional experiencers and creative collaborators, exploring together the vital role and cultural value of artistic practices and communal care. (The text is taken from the work description and the theme exposition of the public safety section of the 4th China Design Exhibition &Public Art Thematic Exhibition)

结　语

自 2012 年 4 月第一次到杨梅竹斜街进行调研至今已有 6 年。在这 6 年中，除了前期的规划、设计以及施工外，设计团队将更多精力投入到了街道改造后的社区营建、居民生活质量提高等方面的探索与实验中。这些近乎公益性质的探索与实验已经远远超出景观设计的范畴，而更多涉及社会学、文化研究以及社会工作的领域中。设计团队从事这些工作的最初动因是想验证设计理念与设计方案是否符合目标受众的需求，以及街道环境的改变对他们的日常生活存在怎样的正面或负面影响。随着调研与反馈工作的展开和近两年来持续的深度介入，越来越多的事实与经验让我们认识到：在多重而复杂的社会问题面前，环境设计所能发挥的作用是极其有限的。

由于北京"老城区改造"工作涉及政府的意志、投资方的利益、老城区原住居民的意愿等多重因素的制约，加之杨梅竹斜街人居环境的特殊性与复杂性，我们所进行的各种实验都只能在微观层面上展开，甚至精确到一户、一人的程度，因此，至今也不能在老城改造与修复、文创产业、居民安置的多重框架内给出一个综合性的结论。究其原因；一是城市现代化进程本身就是各种资源重新配置进而导致利益再分配的过程，其中的利益冲突是不可避免的；二是由于不同社会阶层、群体对"老城文化"的认同与想象之间存在差异，因而存在着规划、设计与"自发秩序"之间的矛盾；三是政府角色的多重性，即投资、文化建设、公共服务之间的内在矛盾。

在当下的主流话语中，这些错综复杂的矛盾与冲突都被认为是中国城市现代化转型过程中不可避免的，只有继续发展才能够获得解决。与此相比，那些声称以人为本的规划与设计理念就略显矫情，甚至虚伪。现代城市规划与设计的根本任务是以城市运转效率为旨归，运用区块功能的分割与连接以求生产效率的最大化。它不仅是对物的规划、设计与改造，同时也是对人的自然状态的机械化规制，将人化约为生产要素，使工作与日常生活分离，而这种分离正是效率最大化的重要保障。所谓的以人为本的"人"也只是在这一秩序规约下的人。便捷的目的是"速达"，休

结　语

闲的目的是更有效率的工作。城市中产阶层理想中的"慢生活"看似是对现代性的抵制，但实则是当下消费社会语境里的一种新型消费模式，是城市生产的一个组成部分。

如今能与这种城市现代化大潮相抵牾的也许只有老城区里普通居民的生活状态了。在经济上，他们虽然属于低收入阶层但有基本保障；在职业上，他们多是或曾经是国企或集体企业的职工；在年龄上，他们多数都已到了退休的年纪或已经退休。他们的生活习惯既有"集体时代"的明显烙印也保留了很多老北京人的习俗。在与他们的接触中，体会最深的就是他们那种闲散、自在的状态，和"吃软不吃硬"的待人处世态度。居住空间的逼仄和经济上的拮据与他们坦然、淡定的日常生活的态度形成了鲜明的反差。为了基本的生活空间，他们不惜僭越"红线"私搭乱建，从杂物堆放到鸽子屋无奇不有，但在十几户甚至几十户的大杂院内仍能保证最低限度的通行空间，这说明在邻里之间还存在着某种有效的协商机制。自家门前、街头巷尾、犄角旮旯就是他们闲聊、下棋的"公共场所"。随手栽种的瓜果爬满房顶，却也不在意邻里采摘，有时甚至是主动地与他人分享。尽管也有为蝇头小利而发生的邻里冲突，但与搬迁、腾退所引发的矛盾相比还是微不足道的。这种自发性的"公共秩序"保证了街道生活的安宁与平静。但是，这种自发秩序的表象就是随意摆放的花盆、粗糙简易

277

的搭建、狭小昏暗的烟酒铺、夏天里赤膊纳凉的男子等各种"脏、乱、差、俗"。规划与设计的任务就是在合理化、规范化、审美化的名义下将这种自发秩序纳入建构秩序的规制之中，以符合城市现代化的整齐划一的功能标准与审美表征。自包豪斯以降，构成主义的形式法则大行其道，名为功能主义的简约风格不仅将世界变成了简单的几何形，其清晰的边界也像刀锋一样将一切混沌的、模糊的、杂乱的事物或切得干干净净，或隐藏其中，或埋入地下。这种在现代艺术史上被誉为"先锋"的审美形式实则与现代化的逻辑具有内在的同构性。在这个原则下，一切自发的、有生命力的和有多重可能性的事物都将被发展的大潮所荡涤。纷繁琐碎的生活经验逐渐被同质化的城市节奏取代，个体和家庭按照自己的习俗自主选择生活方式的可能性越来越少。

北京大栅栏地区老城改造是政府主导的以商业投资模式运作的巨大工程，其中政府扮演着公共服务与投资开发的双重角色。这项工程虽然在名义上是改善老城区的基础设施以及提升老城区居民的生活质量，但实际运作中不免产生投资效益与公共服务之间的矛盾。作为投资者，老城区改造实质上是以挖掘历史文化资源为手段，以开发文化旅游产业为目的；作为公共服务的提供者，这种运作模式较之于单纯的财政投入也许是一个很好的尝试，但如何平衡其

结　语

中的利益关系却难有一个不顾此失彼的完美方案。在我们走访杨梅竹斜街居民的时候听到过这样的感叹："这条街现在变成文化街了，可这和我们有啥关系？"这种抱怨在我们所接触的居民中普遍存在。2017年以后，国家对北京的城市发展策略进行了重大调整，前30年对老城区的大拆大建势头受到了遏制。北京作为首都的政治功能得到了强化，同时，老城保护与修复工作也日益受到重视。在新的城市发展理念策动下，杨梅竹斜街改造与提升项目再次获得了政府的财政支持。与6年前开始的商业开发、居民置换、腾退工程等不同，新一轮改造的重点是古城风貌的修复与文化建设，但执行主体仍然是政府的投资公司。2018年年初，随着国务院机构改革的展开，原有的文化部更名为文化和旅游部。这一名称的改变不仅意味着政府部门职能的调整，也意味着文化不再是一个单纯的意义生产系统，过去相对独立于经济建设的历史、习俗、艺术等将被正当地作为生产要素的组成部分加以优化配置。

在杨梅竹斜街，无论是居民的生活还是房屋及公共设施都遗留着不同时代的各种烙印，包括古代士人的宅邸、民国时代的书局、新中国成立初期的街道工厂以及集体所有制时代的大杂院、改革开放后的发廊、民营旅店以及所剩无几的"传统手工艺人"。在重现古城风貌与文化创意这一主旨下，不同时代叠加的历史遗产将被有选择地重新编排，那些与这一主旨不相匹配的痕迹将被抹去。但是，在我们看来，最不该被抹去的（或许也是最难改变的）是这里普通人的日常生活，无论他们居住于私有的房屋或是公共杂院。

自从杨梅竹斜街改造项目开始以来，这条街道在过去的6年里发生了很大的变化。一方面，不断有新的餐厅、咖啡馆、文艺小店开张或关门，也许这正是文化创意产业的活力所在；另一方面，常驻于此的居民的生活却一如既往，少有可察觉到的明显变化。2017年北京设计周期间，我们曾经策划过将街道居民自家种的丝瓜、葫芦、豆角、辣椒等通过包装设计实现个人或家庭种植品牌化的活动，目的是让文化创意活动也能够惠及街道里的普通居民。但获得的反馈却是：他们都不愿意参与这项活动。多数居民表示他们的种植行为纯属生活习惯，从没想过也不愿参与什么创意或交易的事。从某种角度看，这样的结果也许可以归结为老北京人的惰性，但他们随遇而安的自在性却是一种与现代文明平行的、可互为参照的、实实在在的文化存在。而文化创意产业与过去的传统产业一样，在经济周期、技术升级等因素的影响下，终将被新兴产业替代。但是，只要平凡的生活还在，北京老城区的街道就不会成为文化的"铁锈区"。繁华落尽日，终归平常时。在设计、时尚的浮光掠影表象下，老北京人那种在过往岁月中积淀下来

结　语

的日常生活习性将延绵不绝，代代相承。

由于种种原因，2018 年 9 月，"北京国际设计周"不再以杨梅竹斜街作为展示场地，这条街道也没有了往年的喧嚣，但各家栽种的丝瓜、豆角、葡萄却依然"硕果累累"，就像不曾发生过什么一样。

童岩

2018 年 10 月 于北京

Conclusion

It has been 6 years since the first survey of Yangmeizhu Xiejie was conducted in April 2012. Over the past six years, aside from the previous planning, de-sign and construction, the design team has devoted significantly more energy and resources towards the exploration and experimentation of community construction and the overall improvement of the quality of life of residents after the reconstruction of the street. These explorations and experiments, which are all practically commonweal in nature, have gone far beyond the scope of land-scape design and have more commonality with the fields of sociology, cultural studies and social work. The design team's initial motivation for undertaking this was to test the idea of whether or not design concepts and plans can meet the needs of a target audience or not, and what positive or negative effects may emerge in their daily lives as a result of the changes applied to the street. With the development of greater research and feedback, and continuous in-depth intervention over the

结 语

past two years, the accruement of facts and experiences have made us realize that: in the face of multiple and complex social problems, the role of environmental design is extremely limited.

Due to many restrictive factors involve in Beijing's "old urban renewal" work, such as the will of the government, the interests of the investors and the will of the indigenous residents that reside in the old urban areas, in addition to the particularity and complexity of the living environment present in Yangmeizhu Xiejie, all of the experiments could only be carried out at the micro level, sometimes even being limited in scope to one family or one person. So far, it is impossible to give a comprehensive conclusion based on the multiple frameworks relating to the old city reconstruction and restoration, cultural and creative industries and residential resettlement. The first reason is that the process of urban modernization itself is a process of redistribution of benefits resulting from the reallocation of various resources, in which conflicts of interests are inevitable. Secondly, there are significant differences between different social strata and groups in the recognition and imagination of the "old city culture". Therefore, there is a contradiction between planning, design and "spontaneous order". The third reason relates to the multi-plicity of the role of government, which features inherent contradictions between investment, cultural construction and public service.

In the current mainstream discourse, these intricate contradictions and conflicts are considered inevitable in the process of China's urban modernization and transformation, which can only be solved by continued development. In contrast, those who advocate people-oriented

283

planning and design concepts are hypocritical. The fundamental task of modern urban planning and design is to aim for the highest degree of efficiency in urban operations, by maximizing production efficiency and segmenting and linking block functions. It is not only the planning, design and transformation of things, but also the mechanized regulation of the natural state of human beings, the transfer of human beings into factors of production, and separating work from daily life. It is this separation which is an important guarantee for maximizing efficiency. As a result, these so-called "people-oriented" planners are still operating under this order. The purpose of convenience is to facilitate efficiency, and the purpose of leisure is to enhance the capacity to work more efficiently. The "slow life" enjoyed by the city's middle class appears to be a boycott of modernity, but in reality it is a new consumption model in the current context of consumer society and a part of urban production.

Nowadays, perhaps the only aspect that is in contradiction with this trend of urban modernization is the living conditions of the ordinary residents in old urban areas. Economically, although they belong to low-income groups, they have their basic guarantees; career wise, for the most part they are presently or have been previously employed by state-owned enterprises or collective enterprises, and most of them are old enough to retire or have already retired. Their living habits are marked by both the "collective era" and the long held customs of many old Beijing people. When in contact with them, the deepest im-pression which they convey is their idle, comfortable state, and an attitude of "can be persuaded by reason but not be cowed by force" that they have towards people. The pressures of living space and economic constraints form a sharp contrast with their calm attitude towards daily life. For the sake of basic living space, they are willing to go beyond the

结　语

"red line" to build a private shelter, in a myriad of ways beyond ones imagination, such as constructing a pigeon house from a pile of debris. However, that the limited access to space can be guaranteed in a yard containing multiple or even dozens of families shows that there is still some effective negotiation mechanisms that exist between neighbors. In front of their homes, on the streets, and in hidden away corners, there are "public places" for chatting and playing chess. The fruits and melons that have been casually planted climb all over the roofs, but no one cares about other members of the neighborhood picking them, and sometimes neighbors will even actively share their produce with others. Although there are neighborhood conflicts over small benefits, these are still insignificant compared with those that are caused by relocation and evacuation. This spontaneous public order has ensured the tranquility of their alley life. However, the embodiment of this spontaneous order can be seen in the flowerpots placed freely, the roughshod and simple constructions, narrow and dark tobacco shops, half naked men in the summer, and so on. Essentially, it is seen in all manners of dirty, messy, poor and vulgar. The task of planning and design is to bring this spontaneous order into a form of regulation, which may provide for constructing order in the name of rationalization, standardization and aesthetics, so as to conform to the uniform standards and aesthetic representations belonging to urban modernization. Since the Bauhaus Movement, the formal rule of con-structivism has gradually grown in popularity. This simple style of functionalism not only turned the world into simple geometric forms, but its clear boundaries divide everything with a sharp blade into elements that are chaotic, vague, unclear, hidden, or buried underground. This kind of aesthetic form known as avant-garde is inherently isomorphic in relation to the logic of modernization in contemporary art history. Under this principle, all of the spontaneous things, which feature vitality and a multitude of possibilities, will be swept away by the tide of development. A greater amount of trivial life experiences will gradually become replaced by homogenous urban rhythms, resulting in individuals and families being less likely to choose their own way of life in accordance with their own customs.

The renovation of Dashilan in Beijing's old city area is a government led project, and its operations are based on a commercial investment model, in which the government plays the dual role of both public service guarantor and overseer of investment and development at the same time. Although on paper, the mission of the project is to improve the infrastructure of the old urban areas and improve the quality of life enjoyed by the residents in the old urban areas, the actual operations have inevitably produced a paradoxical situation in which investment benefits and public services are conflicting in nature. As an investor, the transformation of the old urban area is essential so as to tap into the historical and cultural resources as a means to develop the cultural tourism industry for these purposes; as a public service provider, this model of operations may be considered a fair attempt when compared with a purely financial input, however there are significant challenges in finding a way to balance these opposing interests and implement a sound plan regardless of any inherent differences in these models. When we visited the residents of Yangmeizhu Xiejie, we heard a common refrain: "this street has become a cultural street now, but it has nothing to do with us." Such complaints are common among the people we get in touch with. Since 2017, the state has made major adjustments to Beijing's urban development strategy, and the momentum of large-scale demolition and

结 语

construction of old urban areas which had occurred over the previous 30 years has been curbed. In essence, the political function of Beijing as the capital has been strengthened. At the same time, more and more attention has been paid to the protection and restoration of the old city. Driven by this new concept of urban development, the project of the rehabilitation and promotion of Yangmeizhu Xiejie won back the financial support of the government. Unlike the projects that began six years ago, which featured commercial development, residential replacement and the vacation of residents, this new round of transformation is focused on the restoration of the ancient city and cultural construction. However, the primary executive body is still the government's investment company. In early 2018, with the launching of the State Council's institutional reforms, the former Ministry of Culture was renamed the Ministry of Culture and Tourism. This change in title not only signifies the adjustment of the functions of the government department, but also indicates that culture is no longer a production system that is pure in integrity. History, customs and art, which were relatively independent of economic construction in the past, will now be properly allocated as integral parts of the factors of production.

In Yangmeizhu Xiejie, regardless of whether it is people's lives, or housing and public facilities, each and every aspect features the marked remnants of different periods, including the houses of ancient scholars, the bookstores from the Republic of China era, the street factories in the early days of new China, the communal courtyards from the era of collective ownership, the hair salons after opening up and reform, privately-owned motels and a few remaining "traditional handicraftsmen". In accordance with the theme of recreating the ancient city and restoring cultural creativity, these aspects of historical heritage will be selectively rearranged, and any traces that do not match this exact theme will be stamped out. Nevertheless, that which is least deserving of being erased from our view (and perhaps the most difficult) is the daily life of ordinary people, whether they live in private houses or in public courtyards.

Significant changes have taken place in this street over the past six years since the Yangmenzhu Xiejie renovation project first began. On the one hand, new restaurants, cafes and art shops are opening and closing, which very well may be where the vitality of such cultural and creative industries lies. On the other hand, the lives of residents in this place have remained the same, with few noticeable changes. During the 2017 Beijing Design Week last year, we planned to brand the luffa, gourd, beans, and chili peppers planted by street residents themselves by applying creative packaging designs, with the aim of bringing benefit to the ordinary residents in the alley through the application of cultural and creative activities. However, ultimately the feedback was that the residents were reluctant to take part in this activity. Most residents said that farming was purely habitual and that they never considered having any desire to participate in creative or transactional activities. In some ways, this result may be attributed to the inertness of old Beijing people but their easy going attitude exists in a genuine manner as a cultural artifact that is parallel to modern civilization and which can be applied in reference and comparison. Just like traditional industries in the past, cultural and creative industries will eventually be replaced by emerging industries under the influence of economic cycles, technological improvements and other factors. However, as long as ordinary life still exists, the streets of Beijing's old urban areas will never become the

结 语

"rust belt" of the cultural industry. All shall return to normal after bustling flourishing days. Under the superficial skimming surface of so-called design fashion-trends, the traditional daily lives of the Beijingners would continue for generations, just as it had always been like in the past days.

Due to various reasons, Beijing Design Week in September 2018 had no longer sited in Yangmeizhu Xiejie, the street, unlike the past years with exhibition events, had been quiet. The vegetables that the residents had planted, however, still fruitfully with grounds, beans and grapes, like there had never been any changes.

Tong Yan
October 2018, Beijing

2012—2024年杨梅竹斜街环境更新大事记

Highlights from 2012 to 2024: Environmental Renewal of Yangmeizhu Xiejie

2012年4月，杨梅竹斜街沿街立面及节点改造与景观环境提升实施工作启动，2012年9月在北京国际设计周亮相。

2015年7月，杨梅竹斜街66—76号院杂院社区营造工作启动，成立"胡同花草堂"，**居民成为了设计的参与者与实施者**。同年9月，在北京国际设计周期间获得最具人气项目。2015年9月，"相续——北京杨梅竹斜街改造"主题项目参加中国住博会首届城市公共艺术与人居环境展览，并被收录于《2015中国公共艺术年鉴》。2016年1月，该项目入选文化部、深圳市政府共同主办的第二届中国设计大展，并被主办方推荐为优秀项目做专题演讲。

2016年5—11月，以杂院社区营造项目为原型的装置作品《安住·平民花园》入选2016年威尼斯国际建筑双年展中国国家馆，受到设计行业和国内外媒体的广泛关注。作为优秀案例收录于《2016中国公共艺术年鉴》。

2016年9月，北京国际设计周期间，"胡同花草堂"为杨梅竹斜街的居民举办了"杨梅竹花草堂2016——**日常**

生活的景观居民种植展"，展现居民种植经验和平民智慧。

2017 年 9 月，"胡同花草堂"在北京国际设计周期间为杨梅竹斜街居民举办了"杨梅竹花草堂 2017——**'众'瓜得瓜'众'豆得豆**"居民种植展"和杨梅竹斜街夹道社区营造项目"三岁，胡同花草堂"，并建立了引导居民健身的"软组织，胡同中的即时健身系统"和"杨梅竹斜街小气候监测数据交互展"。

2017 年 11 月，杨梅竹斜街社区治理与公共空间提升项目启动。提出系统性的设计方案解决现状诸多问题。

2018 年 9 月，"胡同花草堂"在北京国际设计周期间为杨梅竹斜街居民举办了"杨梅竹花草堂 2018——**繁华落尽终归平常**居民种植展"。

2019 年 1 月，《胡同花草堂》作品入选在深圳市当代艺术与城市规划馆举办的由文化和旅游部等部门共同主办的第 3 届中国设计大展及公共艺术专题展。

2021年4月，受国际建筑师协会邀请，无界景观工作室主持设计师谢晓英在巴西里约热内卢第27届世界建筑师大会（UIA2021）以《安住》为主题、杨梅竹斜街有机更新为案例进行演讲。同年，杨梅竹项目入选联合国亚太城市可持续发展目标优秀项目库。

2021年9月，杨梅竹斜街"胡同花草堂"与埃塞俄比亚谢格尔公园"都市花草堂"进行现场连线活动，让"一带一路"国家的普通百姓互相了解，建立友好连接。

2021年10月，《抗体·风景融入日常生活——即时健身系统》入选在深圳市当代艺术与城市规划馆举办的由文化和旅游部等部门共同主办的第4届中国设计大展及公共艺术专题展。

2023年3月，杨梅竹斜街《胡同花草堂》入选在北京世纪坛举办的由中国美术学院美丽中国研究院主办的"2023美丽中国纪事"展。

2023年，杨梅竹斜街"胡同花草堂"作为北京中轴线上重要的"一带一路"民间文化交流空间，记录于北京中轴线主题国际传播系列纪录片《京之轴》。

2012—2024年杨梅竹斜街环境更新大事记

2021年，受国际建筑师协会邀请，无界景观工作室主持设计师谢晓英在巴西里约热内卢第27届世界建筑师大会（UIA2021）上，以《安住》为主题、杨梅竹斜街有机更新为案例进行演讲。同年，杨梅竹项目入选联合国亚太城市可持续发展目标优秀项目库（SDGs）。

In 2021, our chief designer at View Unlimited Landscape Architects Studio, Xie Xiaoying, was invited to speak at the 27th World Congress of Architects (UIA 2021 RIO) in Rio de Janeiro, Brazil, where she showcased our "HOME" theme and the organic renewal of Yangmeizhu Xiejie. That same year, the Yangmeizhu project was featured in the United Nations Urban SDG Knowledge Platform.

293

2019年初，无界景观团队《胡同花草堂》作品入选在深圳当代艺术与城市规划馆举办的第三届中国设计大展及公共艺术专题展。中国美术家协会主席、中央美术学院院长（前中国美术馆馆长）范迪安先生与多位专家学者、政府官员参观了展览，上海大学上海美术学院副院长金江波教授对该作品进行了现场讲解。

In early 2019, the work "Hutong Flora Cottage" by the View Unlimited Landscape Architects team was selected for the Third China Design Exhibition & Public Art Thematic Exhibition held at the Shenzhen Museum of Contemporary Art and Urban Planning. Mr. Fan Di'an, Chairman of the China Artists Association and President of the Central Academy of Fine Arts (former Director of the National Art Museum of China), along with several experts, scholars, and government officials, visited the exhibition. Professor Jin Jiangbo, Vice Dean of the Shanghai Academy of Fine Arts at Shanghai University, provided on-site commentary on the artwork.

2012—2024年杨梅竹斜街环境更新大事记

2015年10月,荷兰国王威廉·亚历山大(Willem Alexander)参访杨梅竹斜街(左图)
2024年3月,荷兰首相马克·吕特(Mark Rutte)访问中国期间参观杨梅竹斜街(右图)
In October 2015, King Willem-Alexander of the Netherlands visited Yangmeizhu Xiejie (left picture)
In March 2024, during his visit to China, Prime Minister Mark Rutte of the Netherlands visited Yangmeizhu Xiejie (right picture)

2024年6月,俄罗斯建筑师联盟主席尼古拉·舒马科夫(Nikolay Shumakov)一行6人参观杨梅竹斜街
In June 2024, a group of six people including Nikolay Shumakov, President of the Union of Architects of Russia, visited Yangmeizhu Xiejie

安住·杨梅竹斜街改造纪实与背后的思考

2019年雪后的杨梅竹斜街街景（摄影：马志骜）
Street view of Yangmeizhu Xiejie after a snowfall in 2019 (Photo by Ma zhiao)

项目团队
Project Team

杨梅竹斜街环境更新（2012—2013）：

中国城市建设研究院无界景观工作室：
谢晓英、张琦、瞿志、王欣、张元、邹雪梅、欧阳煜、陶陶、吴悦、冀萧曼、曾胡英、王翔、孟庆诚、吴寅飞
场域建筑：梁井宇、叶思宇等
原研哉设计院
八股歌互动
大栅栏投资有限公司：徐正榕、李紫祥、葛文彤、贾蓉、吴奇兵、王建伟

杨梅竹斜街杂院夹道社区营造（2015）：

中国城市建设研究院无界景观工作室：
谢晓英、童岩、黄海涛、周欣萌、张元、王欣、邹雪梅、鹿璐、张婷、瞿志、李薇、李萍、段佳佳、李银泊
中国建筑设计研究院：陈一峰、杨晓东等
杂院居民：王秀仁、张修良、东雪梅、魏兰涛及家人、暂住保安等
大栅栏跨界中心：贾蓉、姜岑

第15届威尼斯国际双年展参展作品（2016）：

中国城市建设研究院无界景观工作室：
谢晓英、童岩、黄海涛、瞿志、周欣萌、张琦、张元、鹿璐、冀萧曼、邹雪梅、李萍、王翔、雷旭华、王欣、高博翰、杨灏、吴寅飞、刘旭、张婷、吴悦、李薇、孟庆诚、段佳佳、李银泊、陈宇冰
杂院居民：王秀仁、张修良、东雪梅、魏兰涛及家人、暂住保安等

北京国际设计周展览（2016—2018）：

2016
中国城市建设研究院无界景观工作室：
童岩、张元、吴寅飞、刘旭、邹雪梅、吴迪、段佳佳
北方工业大学：杨鑫教授等
居民：段宝玺、东雪梅、魏兰涛及家人、高魁、赵琴、孙淑华
大栅栏跨界中心：贾蓉、姜岑、张雅萌等

2017
中国城市建设研究院无界景观工作室：
童岩、黄海涛、谢晓英、张琦、张元、鹿璐、邹雪梅、李萍、高博翰、曲浩、田南、王欣、段佳佳、王翔、吴寅飞、李银泊、刘旭、李宗睿、苏争荣
北方工业大学：杨鑫教授等
大栅栏跨界中心：贾蓉、姜岑、李美玲等
居民：高魁、赵琴、孙淑华、吴素鑫、王建民、芦殿珍、王文湘、马栓海

2018
中国城市建设研究院无界景观工作室：
谢晓英、童岩、张元、王翔、吴迪、王欣、李萍、李宗睿、段佳佳、田艾、曲浩、吴寅飞、刘旭、田南、邹雪梅、苏锦钒
居民：高魁、赵琴、孙淑华、吴素鑫、王建民、马栓海、王文湘、张英、刘广全

杨梅竹斜街社区治理与公共空间提升（2018）：

中国城市建设研究院无界景观工作室：
谢晓英、黄海涛、张元、周欣萌、张琦、王欣、田艾、李银泊、李宗睿、高博翰、杨灏、段佳佳、曲浩、吴迪、王翔、吴寅飞、邹雪梅、孙莉、苏锦钒、贾蓉、姜岑
北京工业大学：熊文教授　大料建筑：刘阳等
大栅栏投资有限公司：徐正榕、葛文彤、李紫祥、宋苗、王建伟

中埃文化交流活动（2019-2024）：

中国城市建设研究院无界景观工作室
Etoile General Contractor PLC,ETHIOPIA：Mr.Kaleb Menwyelet Munacha
商户：采瓷坊、斜街8号咖啡馆
CityLinX 设计联城创始人、首席执行官：贾蓉
大栅栏投资有限公司：宋苗、滑斌
居民：高魁、赵琴、王秀仁、张修良、吴素鑫、段宝玺、魏兰涛等

无界景观设计工作室

View Unlimited Landscape Architects Studio, CUCD, CCTC

无界景观工作室是中国建设科技集团中国城市建设研究院的风景园林专家工作室，主持设计师为谢晓英。我们致力于从绿色发展角度出发，在构建国家自然生命支持系统、基础设施化的城乡绿色空间和绿色化的市政工程基础设施等层面，探索景观设计统筹多专业协作的最大可能性，助力推进生态文明建设。我们致力于探寻因地、因时、因人而异的，与同质化相悖的解决方案，为区域、城市乃至国家树立形象标识，增强文化认同，提升民心凝聚力与民众自豪感。我们关注日常生活与公共空间的联系，关注景观设计对社会关系的良性引导，以专业手段协调人与环境的关系，缓解人的生存压力，激发民众活力与场地生产力，提升"安住"者的幸福感与归属感。将风景融入日常生活，为城市与乡村营建整体的、连续的美丽，为绿色可持续发展提供从实践出发的有效助力。

View Unlimited Landscape Architects is affiliated to China Urban Construction Design & Research Academy Institute. Lead by landscape architect Xie Xiaoying, We are dedicated to exploring the greatest possibility for landscape design to serve as the main coordinator in multi-disciplinary collaboration. From the standpoint of green development, this includes building the national natural life support system, the infrastructural development of urban and rural green space, and the green transformation of municipal engineering infrastructure, all contributing to the progress of ecological civilization.We are committed to identifying solutions that veer away from homogenization by taking into account the surrounding environment, time period, and community, aiming to establish iconic symbols for regions, cities, and the country, to enhance cultural identity, and to improve public unity and pride.We focus on the connections between daily life and public spaces, steering society towards a positive direction through landscape design. Using our professional expertise, we aim to balance the relationship between people and the environment. This alleviates the pressure of human survival, stimulates public vitality and the productivity of places, and enhances the sense of happiness and belonging for the inhabitants. By integrating scenery into everyday life, we strive to create holistic and continuous beautification in urban and rural spaces, providing effective support for green sustainable development through our practices.

致　谢
Acknowledgments

胡军、王雪语、王德厚、赵园、吕大年、黄梅、吴文一、
梁井宇、申献国、徐正蓉、葛文彤、贾蓉、李紫祥、宋苗、
李志斌、张玉清、张立新、迟义寰、谢爱伟、吴士新、
丁高、王学东、张富友、王磐岩、王香春、张琰、毕婧、
周伯任、马晓暐、湛旭华、杨敏、孙伟雁、杜云录、孙妤、
童曦、张晓佳、李然、竹倩、齐欣、耿毅军、黄颖、
康涛、陈庆宗、胡长青、赵昭、陈一峰、徐坤、朱红星

史建、王明贤、崔愷、谢小凡、诸迪、范迪安、郑浩、
谭平、李虎、赵心舒、曹玥、闫东、刘振林、解严一、
杨欣、李芸芸、Wang Yile、Daniel Lenk

宁波万盛投资有限公司：刘世南
中国城市建设研究院有限公司
北京泛亚照明装饰工程有限公司：柴志伟
北京新枫岚园林绿化工程有限公司：汪斌峰
北京正和诚文化发展有限公司：温全、字云、曹浪洲
北京市规委宣教中心主任：刘俊兰
北京凯欣城市发展咨询有限公司总规划师：张硕辰
Etoile General Contractor PLC,ETHIOPIA：
Mr.Kaleb Menwyelet Munacha

致　谢

中国文化报：李亦奕
中国新闻社：邓苗、杨柳
中国网：王志永
北京青年报：赵晓笠
中国美术报：王雯雯
新京报：陈乱乱
中国出版促进会：陈垠、李中南

住区：王若曦
LAF 景观设计学：佘依爽
风景园林：侯晓蕾、郭巍
中国公共艺术年鉴：王永刚、卢远良
公共艺术：吴蔚
艺术中国：刘鹏飞

上海大学上海美术学院：金江波
南通大学艺术学院：孙婷
《京之轴》摄制组：李然、王佐、张小蓓、王文超
美丽中国研究院策展团队：汪莎、张春艳、吕珮瑄、俞洲、周嘉鸿、刘艺璇、徐欣怡、唐麟、张真予、王瑾雯、卢意

感谢十余年来所有对杨梅竹关注支持的各位朋友及杨梅竹斜街的居民们